IEEE ENGINEERS GUIDE TO BUSINESS

Volume VII

STARTING A HIGH-TECH COMPANY

by

Michael L. Baird

The Institute of Electrical and Electronics Engineers, Inc.
New York, New York

IEEE *Engineers Guide to Business Series*

Editor: Barbara Coburn
Typographer: Jill R. Cals
Cover Design: David Beverage

Printed in the United States of America

Library of Congress Cataloging-in-Publication Data

Baird, Michael L.
Starting a high tech company / by Michael L. Baird.
p. cm. -- (IEEE engineers guide to business ; v. 7)
ISBN 0-7803-2293-2
1. High technology industries--Management. 2. Electronic industries--Management. 3. New business enterprises--Management. I. Institute of Electrical and Electronics Engineers. II. Title. III. Series: IEEE engineer's guide to business ; v. 7.
HD62.37.B354 1995
620'.0068--dc20 95-16795
CIP

CONTENTS

SIDEBARS

FOREWORD

Entrepreneurship is the call of the 1990s. This is particularly the case in high technology, where our society continues to place more feverish demands on American ingenuity for improvement and innovation in the products and services that surround our lives. Everyone looks to create the fastest chip, the lightest laptop, and the most powerful software. There is no limit to the technological advances realized with every new product introduced, which is exactly why there has never been a better time to consider an entrepreneurial venture.

It is hard to escape the image of entrepreneurial success in today's media-conscious environment. Multi-billionaire founder Bill Gates of Microsoft has appeared repeatedly on the covers of business magazines. Entrepreneur Ross Perot reaches out to be the president of the United States. Successful entrepreneurs are ubiquitous at a time when loyalty to *Fortune* 500 companies continues to diminish amid the trend of layoffs and reorganization. When you consider that the overwhelming majority of new jobs created this decade will come from start-up companies, a high-tech business venture seems a reasonable, if not downright admirable, proposition.

As the president and chief executive officer of Silicon Valley Bank, I have met with thousands of ambitious entrepreneurs from every walk of life. I am very proud to have provided guidance and assistance to many of these individuals, many of whose companies are now quite successful. What do these successful individuals all have in common with each other?

You will not find it surprising when I tell you that these individuals all share the same trait: motivation. Not motivation for wealth, money, or power, though. Rather, the more successful people are those who desperately want and seek out autonomy and challenge. They want to use their drive, skills, and hard work to turn an idea or a dream into a company. Start-up entrepreneurs have a special passion for life and view each new workday as a new challenge. Of course, I know my generalities cannot assimilate life in a high-tech start-up, but I believe *Starting a High-Tech Company* delivers the important information and sage advice those of you considering a high-tech venture need in order to make informed decisions on how to proceed.

Many entrepreneurs who want to start a venture do not know where to begin or what to expect. In this book, classic formulas for success are laid out in no-nonsense terms. Life in a start-up is described vividly, and real-world anecdotes detail both failures and successes. Valuable advice is presented for creating wealth using founder's stock and stock options, which are topics not always familiar to technically oriented entrepreneurs. While the book specifically addresses the high-tech entrepreneur, those considering employment at a start-up will find it helpful in understanding the dynamics involved in working in that environment.

In the investment community, we see higher quality start-ups today than in the 1980s. Competition for venture funding is intense and global. Successful companies plan from day one to attract syndicates of investors who will carry the company from the concept stage through an initial public offering or acquisition by a larger, usually public, company. Many start-ups establish early strategic partnering relationships with customers, suppliers, and governments worldwide. Those companies that are well planned, are created by strong teams, and offer proprietary technology that plays into growing and targeted markets can obtain funding. With funding comes an excellent chance for success. This book will arm you with the knowledge necessary to create such a company.

Life in a start-up is one of the most exciting things you can experience, and I encourage you to explore your options. Do your homework, find your place in life, and live your entrepreneurial dream.

Roger V. Smith
Silicon Valley Bank
April, 1995

PREFACE

Career stability is becoming increasingly volatile. For the most part, the business re-engineering process has cast off the livelihood of the individual as a corporate concern. For most professionals, the once comforting career track has become obsolete. Workers must now re-invent themselves—and increasingly, this takes the form of launching a new business. For most of you, starting your own high tech company has long been a dream—it may now be a necessity. Today's job pressure can be the catalyst for your start-up success.

During the past decade I have had the opportunity to interact with a large number of high-tech entrepreneurs at all stages of development. It seems that the desire for men and women to achieve independence and free themselves from the bonds of traditional employment is more deeply ingrained in the human soul and more universal in scope than commonly thought. If you are such an individual— if you hold in your soul a powerful ambition to start your own high-tech business, now is time to act.

After enjoyable and profitable experiences as an officer in six venture capital and other privately funded turnarounds and start-ups, I remain involved almost exclusively with start-ups. Because so many people have asked for my advice over the years, I committed to put it into writing. I hope that you too will benefit financially and enjoy creating your new business.

In preparing this book, I have made a special attempt to verify all information, especially as it pertains to the subjects of law, securities, taxes, and investments. However, this information changes frequently, and varies by state. Therefore, on these important investment, legal, and tax matters, follow this book's advice by seeking expert counsel. You must be prepared to assume full responsibility for the outcome of your decisions.

One thing is certain: With your common sense and this book as a guide, you can determine the risks and rewards of starting a new venture. This book tells you which questions to ask and what information you need to obtain to make an informed decision.

I can guarantee you that starting your own business is going to be the greatest thrill of your life. Creating value for your customers, creating jobs for your employees, contributing to your own success, and giving yourself and your family complete psychological and financial independence will make all the work worthwhile.

Read and enjoy!

Michael L. Baird
April, 1995

ACKNOWLEDGMENTS

"When life deals us blows that we can't overlook, some suffer in silence and some write a book."
—E. B. de Vito, *Wall Street Journal*, July 7,1987

Completing this book was a wonderfully satisfying affair, and I have lots of people to thank for helping to make it happen. I especially want to acknowledge the gift of precious time and the many contributions made by the following reviewers: Janet Brewer, attorney-at-law with the Law Offices of Janet L. Brewer in Palo Alto, CA; Kenneth R. Allen, attorney-at-law with Townsend and Townsend in Palo Alto, CA; Janet G. Effland, vice president of the venture capital firm Alan Patricof Associates, Inc., in Menlo Park, CA; Anthony C. Bonora, senior vice president of research and development at Asyst Technologies in Milpitas, CA; the late David H. Bowen, publisher of *Software Success* in San Jose, CA; William J. Wall, vice president of finance and administration and CFO of Resumix, Inc., in Santa Clara, CA; Dr. David K. Lam, founder of Lam Research Corporation and president and CEO of Expert Edge Corporation in Palo Alto, CA; Ed Zschau, former U.S. congressman and chairman and CEO of Censtor in San Jose, CA; Dr. Phillip B. Nelson, industrial psychologist with the Institute for Exceptional Performance in San Francisco, CA; and Dr. Jeanne Gilkey of the University of Phoenix in San Jose, CA.

In addition, the following individuals provided advice or resource materials: Bruce W. Jenett, attorney-at-law with Fenwick & West, Palo Alto, CA; Dr. Jim Plummer, president of Q.E.D. Research and venture capital consultant in Palo Alto, CA; C. Gordon Bell of Los Altos, CA, formerly of Digital Equipment Corporation and author of *High-Tech Ventures;* Mary Cole; and Kathy Janoff.

My wife Heidi grammar-checked the manuscript. My terrific kids, Robby and Sandy, missed way too many bike rides with their dad.

Barbara Coburn, Manager, Career Development Programs at IEEE, helped me step-by-step through the arduous process of revising and marketing the book. Jill Cals, Editorial Coordinator orchestrated production of this book. David Beverage designed the cover, and Jill Cals typeset the book.

INTRODUCTION

The fact that start-up companies have been generating new jobs while large established companies have been laying people off has caused a great deal of media attention to be focused on start-ups. At the same time, starting your own business has been glamorized by a steady stream of articles and books about successful entrepreneurs.

The good news is that all of this exposure has made it socially acceptable for people to quit their big company jobs and start their own enterprises. The bad news is that much of the available information for new companies glosses over or completely ignores the monumental effort required to successfully launch an organization. Over the years, numerous entrepreneurs have told me about the discouragement and hard work they faced in launching their companies.

Starting a company is a very difficult process, but much of the discouragement results from unrealistic expectations. When I left Amdahl Corporation to launch my first software company, my boss at the time told me, "Dave, you think this is going to be a sprint, but it's really a marathon." He was the first of many wonderful people who extended a helping hand and showed me the lay of the land.

Starting a company in Silicon Valley enabled me to meet other entrepreneurs who were further along in their ventures. Some had started dozens of businesses, while others were just beginning their second year. No matter what stage of the game they were at, I learned something from everyone I talked to.

While reading Starting a High-Tech Company, I thought back to my early days as an entrepreneur and gleaned new insights into issues I had wrestled with time and time again. Mike Baird reached out and helped me, and, in turn, I shared some of my experiences with Mike, which are included in this book.

If you are considering initiating your own start-up from a large-company environment, I strongly urge you to read this book from cover to cover.

Then, reread it every few months and refer back to it with questions as they come up in your business.

If you have already started your own business, much of what you read here will confirm what you already know, but you may gain deeper insight into why things are the way they are.

Everyone who reads this book will feel a sense of camaraderie as they read about other people who have embarked on this journey.

Good luck to you all as your businesses grow and evolve!

David H. Bowen

Part One

THE GENESIS

Part One of *Starting a High-Tech Company* discusses opportunities for the entrepreneurial engineer, clarifies your new role as the CEO and founder of your own business, and describes life in your new start-up. If you have not given much thought to how starting your own business might impact your life, then these chapters should be especially beneficial.

Chapter 1
START-UP OPPORTUNITIES FOR HIGH-TECH ENTREPRENEURS

"Software opportunities continue to provide the best paths for engineers wanting to start their own companies."
— David H. Bowen

OPPORTUNITIES FOR START-UPS ABOUND

While the number of start-ups may be down from its highest levels, the number that will succeed is not. Although it is true that less venture capital is available these days, those companies good enough to get it are more likely to do well.

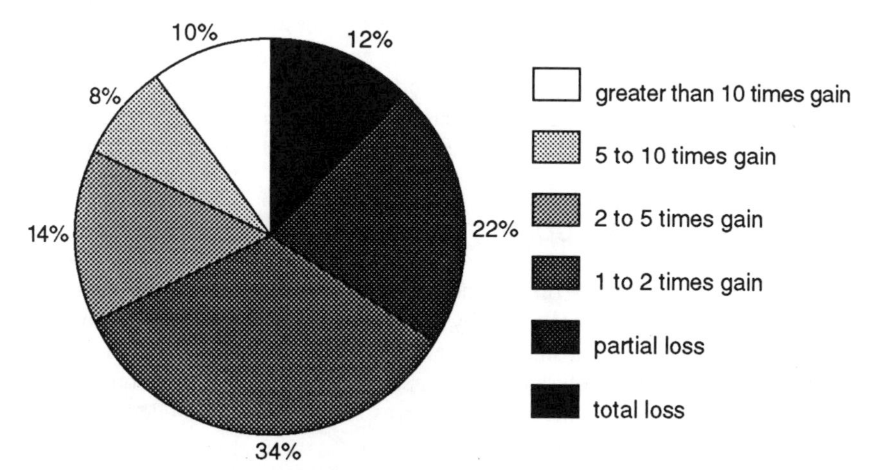

Figure 1.1 Rates of Return for 200 Venture Capital-Backed Ventures from 1973—1983
Source: TTG Research and J. Trudel, *High Tech with Low Risk: Venturing Safely Into the 90s*

As Figure 1.1 suggests, venture capital-backed companies have made money for some, but not all, start-up entrepreneurs. While a venture fund's investors have the safety of diversification, your shot must be on target.

Your gain will roughly mirror that of your investors' return in your start-up. So you see, you have roughly a one-third chance of losing money, a one-third chance of breaking even, and a one-third chance of becoming substantially wealthy.

This book teaches you how to launch and finance your successful start-up. There remain numerous areas of opportunity for engineering-related start-up companies, some of which involve markets far larger than those of the 1980s. These new technologies will continue to drive out the old. For example, personal computers are increasingly becoming connected in networks and supplanting what minicomputers and mainframes used to do. That requires new hardware and software tools that you could develop. Hardware and software for pen-based computers are hot, as is software for WindowsTM-based applications. Medical electronics and biotechnology also continue to cross new horizons, and wireless communication is emerging as an important growth area.

However, as you will come to appreciate, your business success will depend on much more than simply developing an exciting new technology. But that is what you are best at—developing new technology. While others may be expert in marketing, finance, and other aspects of business, they will lack your knowledge of key enabling technology. So who will be the winner? You are betting that you can learn how to plan and build a successful technology-based business faster and better than a nontechnical businessperson can learn how to exploit technological know-how. To discover whether that is a good bet, read on. The winner is usually a person who is most determined to win, and it seems that you have a good head start.

THE LURE OF FREEDOM

Autonomy clearly ranks first on the list of reasons why individuals start their own companies, followed by the desire for income and wealth. Figure 1.2, adapted from studies of new firms in Minnesota conducted by Paul D. Reynolds, professor of business administration at Marquette University, illustrates some of the major reasons cited by entrepreneurs for wanting to start their own businesses.

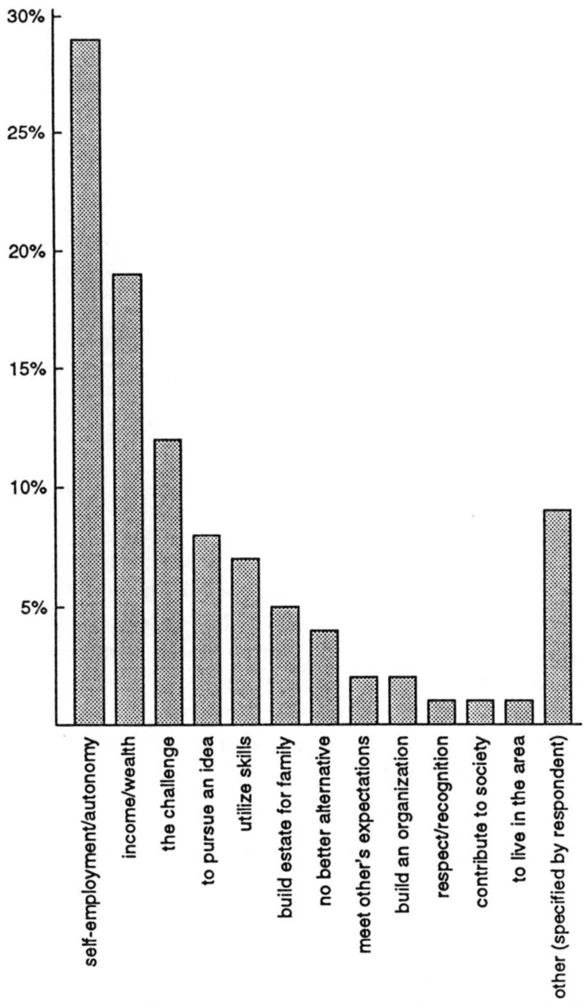

Figure 1.2 Reasons Cited for Starting One's Own Business

THE PROFESSIONAL ENGINEER

There are over 5 million professional engineers and scientists in the United States. These well-educated, hardworking men and women represent some of the best and brightest talent in our country. Their contributions to the profits of business and industry represent billions of dollars each year. Yet many of these individuals will work extremely long hours, often for minimal satisfaction, security, and financial reward.

One MIT study indicates that about half of these professionals have seriously considered starting their own businesses. Another study by Execunet determined that 60% of corporate executives would start their own business if they could. Sixty percent also would prefer working for a smaller company if they changed jobs. Your dreams are not alone, and this book is the answer. It will teach you how to launch your own successful business and accumulate significant wealth in the process.

In the remainder of this book I often use *engineers* as a generic term that applies to many other technology-oriented professionals. For example, if you are a research scientist wanting to start a business, you will need to become more applied and less theoretical. To be successful, entrepreneurial engineers must become business planners and marketers as well.

THE RECENT COLLEGE GRADUATE

About 200,000 engineering and science college students graduate each year. Many of them dream of starting their own businesses, but are especially frustrated by lack of experience. Nevertheless, this may be the best time to start your own business. Think about it—you have unequaled energy, enthusiasm, fresh knowledge, and university contacts, all of which will diminish over time. Many people at this point in life think that they can do anything, and often they are right! Family burden is frequently less of a problem for a fresh graduate also. What are the risks of taking a plunge?

FROM TECHNOLOGY TO PRODUCT TO MARKETING

Whether you are a practicing engineer or a recent college graduate, if you have an interest in starting your own business there are many opportunities for you to explore. Dreaming about it is not enough. You need to plan a course of action to launch your successful technology-based business.

- Identify appropriate products to develop based on your technological skills. (As will be discussed in this book, while your products will be technology-based, your business must be market- and customer-driven and technology fueled.)
- Determine how to develop and produce those products.
- Take your products to market rapidly and successfully.

IS IT TIME TO CREATE YOUR OWN JOB?

With a couple of tenuous exceptions such as IBM or Hewlett-Packard, there really is no such thing as job security, even in a *Fortune* 500 company. IBM, for example, is making it tougher for those seeking lifetime employment through their implementation of a new policy requiring a substantial percentage of employees to be rated "unsatisfactory" in performance reviews—forcing many to leave.

Since the mid-1980s, as corporations have responded to global competition and technological change by merging and consolidating, downsizing and de-layering, 2 million middle-management positions have been permanently eliminated. American corporations have unilaterally repealed the unwritten law that once bound them to their managers, and have been jettisoning them in carload lots.

The U.S. Bureau of Labor Statistics assistant commissioner Martin Ziegler says that statistical revisions will add 650,000 more jobs lost in the 1991 recession to bring the total number to more than 1.4 million. Furthermore, 700,000 more jobs will be lost in 1992 if corporations keep slashing payrolls at their current pace, says Dan Lacey, editor of *Workplace Trends*. Could your job be one of these?

According to the Bureau of Labor Statistics, early in 1992, the official unemployment rate was around 7.1%, representing 8.9 million Americans. Estimates of true unemployment, however (factoring in the 6.3 to 6.7 million workers holding part-time jobs because that is all they can find and the 1.1 million discouraged workers who are no longer being counted), ranged around 10.4%. In 1991, one in every five American workers was unemployed at some point (25 million people, almost 20% of the workforce) according to the Conference Board, a business research center in New York. One in every four U.S. households in early 1992 included someone who was unemployed in 1991.

David Bowen, publisher of *Software Success*, points out that the real problem for a middle-aged middle manager is holding his or her job for another 20 to 30 years:

> Middle managers are defined as making over $40,000 per year. About one million middle managers lost their jobs in 1991. Since there is about a ten percent chance of losing your job every year, over 30 years, there is only about a four percent chance of holding your job! $(0.90)^{30} = 4.24\%$. And, middle managers who change companies every five to ten years will end up without good retirement plans since they never stayed put.

ISSUES TO CONSIDER

Quitting your job and starting a company is stressful and full of uncertainty. If you are a typical reader, you have been employed for several years in a large, stable company. If you are seriously considering leaving a position that has the appearance of security and a good salary (although with limited financial upside) for the excitement of the fast lane in a start-up, you need to consider for more than a moment what this means to you and your family. There are many important issues, and you need facts to satisfy your concerns. I have listed some major questions that you need to ponder as you read the remainder of this book.

- What are your life goals?
- What are you getting into, and is this really what you want to do? Are you prepared for very hard work, or are you more of a "quality-of-life" person?
- Will your business have a chance to succeed financially? Are you willing to bet your chances for success with one or two other key employees?
- What is your quality of life now, and how would it change?
- Can you separate the excitement and glamour of a start-up from its reality?
- Are you prepared to be consumed by your business? It will never let up and you will never escape it during its formative years.
- What can a start-up do to you physically and mentally? Are you strong and healthy enough to pull off a start-up?
- What are the time demands of a start-up? Do you like to recreate on weekends, or will you work? How much time do you want or need with your family?}
- Are you ready for extensive travel and "give it all you have" performances for customers and investors?
- Does establishing and maintaining a reputation in, for example, the research community mean a lot to your personal development? Is going to technical conferences important? Will a start-up afford such luxuries?
- Will you escape that *Fortune* 500 feeling of being a wage slave even if you launch a start-up? Are there other types of captivity that will trap you?
- Realistically, what is the chance to become independently wealthy?
- Can you survive without a paycheck for three to nine months, either while your start-up is getting funded or after your start-up crashes and burns?

- What are the alternatives if you stay? Is there something better between a start-up and your current employer?
- Have you considered the possibility of a start-up ruining a stable marriage?
- Do you thrive on continuous change (not always improvement) or despise it?
- How old are you? When is the best time to move?
- Last, and perhaps most important, is your spouse and family. Will they be fully supportive and excited as well? If not, the additional stress makes your odds much worse. Will you have their support? Their support is critical, since they will share with you the inevitable financial and time sacrifices.

SMALL BUSINESS: NOT SYNONYMOUS WITH START-UP BUSINESS

The words *small business* and *start-up* at first may seem synonymous. Clearly not all small businesses are start-ups (check out the VCR rental store on the corner), but most start-ups do begin small. Your start-up is the result of setting in motion a new company, and to what size and at what rate your company grows is critical to your financial success. Though there is some debate concerning what constitutes a small business, for the purposes of this book it is defined as an independently owned and operated company with fewer than 20 employees. A company expanding at a rate of more than 10% a year is considered fast-growing.

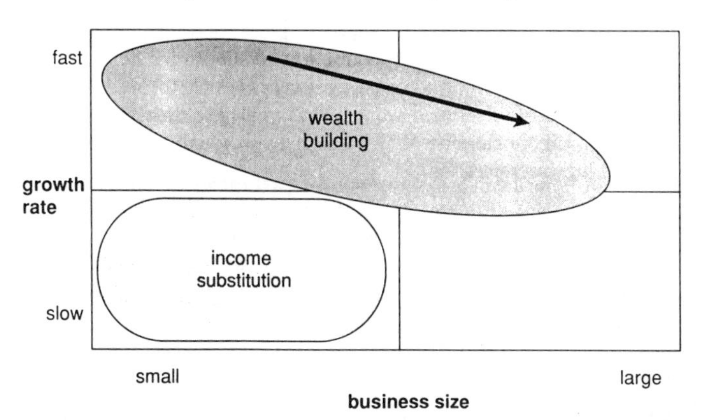

Figure 1.3 The Income Substitution—WealthCreation Spectrum

Your start-up business, if it survives, is destined to become either an income substitution business or a wealth-building business, as illustrated in Figure 1.3, depending mostly on how fast and large it grows.

People who simply do not want to work for someone else can easily start up a small income substitution business, such as a one-man lawn mowing service. Joe may even earn as much mowing lawns as he did working for Mr. Bemis, and this may make Joe happy. This kind of small business is called an *income substitution business.*

A *consultancy* is a business formed by an individual to provide services and is generally limited to creating an income stream. A *proprietorship* is a business formed by an individual or related family members and also is generally limited to income substitution. You will often find unemployed engineers holding themselves out as consultants.

If your small business does not have high growth as an objective, and it is not team-driven, it most likely will not become the wealth-building vehicle you need for financial success and independence (see Figure 1.4). The successful wealth-building start-up business for engineers, for which this book was written, was characterized by the charter of the bygone Silicon Valley Entrepreneurs Club, which assisted entrepreneurs in creating and managing team-driven and high-growth companies. Such companies commonly have annual sales goals of from $10—$100 million or more over a period of three to five years.

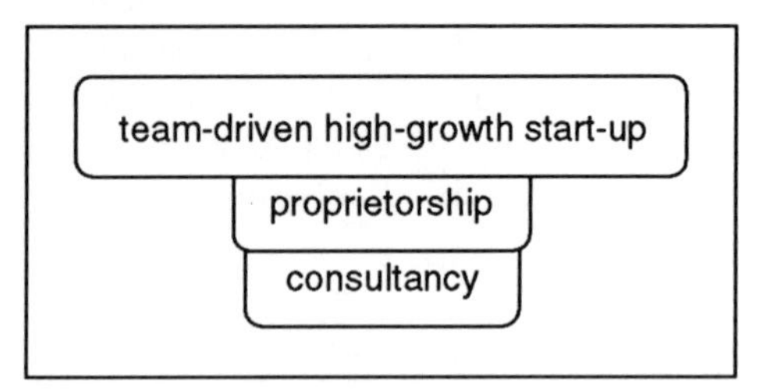

Figure 1.4 Forms of New Engineering-Related Businesses

If Joe has the makings of a true entrepreneur, there is nothing to prevent him from growing a very profitable lawn mowing organization that could create great wealth for him. There may be a need for a nationally recognized quality lawn care franchise, for example.

Robert Ronstadt, publisher of *Ronstadt's Financials*, a software package for financial planning, similarly categorizes ventures into three categories:

Type of Venture	Sales Range	Employees
lifestyle	0 to $1 million	0 to 4
smaller high-profit	$1 million to $20 million	5 to 50
high-growth	over $20 million	over 50

Individuals who want independence and autonomy start lifestyle ventures. These people do not want the aggravation that growing a business entails, and they prefer to conduct their business lives much like their personal lives. There is nothing wrong with wanting to conduct a lifestyle venture, but if that is really your objective, do not try to act like a venture-backed start-up.

Smaller, high-profit ventures allow the entrepreneur not to relinquish equity or ownership control. This can be a challenge to an engineer who may not have access to the financing that a high-growth business requires. This book will be useful for smaller, high-profit business venture planning.

Successful high-growth ventures usually lead to nationally and internationally known businesses. Significant outside funding is required to grow this kind of venture. The engineer aspiring to create a high-growth business will be seeking to maximize the market value of the company, and in the process will create significant wealth for himself or herself, the venture's investors, and many of the company's employees.

TAKE RISKS!

One study from UCLA suggested that in kindergarten, 25 percent of students show a natural need for high achievement and a willingness to take risks. By the time they get to high school, only three percent do.

If you really do have that entrepreneurial craving, do not hang on to a job you do not love—jump on the start-up bandwagon, take some calculated risks, and enjoy the rest of your life.

It would be great to retire at 40, if that is what you want to do, but I only know of four ways to such financial independence, and only the last one is truly satisfying and in your control:

- You might marry it.
- You might inherit it.

- You could steal it.
- You could earn it in your start-up.

If risk-taking is not part of your personality, however, you may want to seriously consider keeping your *Fortune* 500 job for as long as possible. Bowen, advisor to many would-be entrepreneurs, says, "I talk to many 'big company employees' who think they want to take risks, but they don't have a clue what risk really is." Look into yourself and try to see what is inside. What is right for you?

Resources Available to Start-up Entrepreneurs

ACCOUNTING FIRMS

Many of the top accounting and consulting firms along with many banks and law firms publish excellent, free booklets on venture capital, securing financing, private placements, writing business plans, growing a business, business valuations, doing business overseas, going public, etc. By simply making a few phone calls from the Yellow Pages you will have more quality reference material than you will ever need. Some good places to start are:

Coopers & Lybrand

(Ask for *Charting a Course for Corporate Venture Capital* and *Valuation Services.*) Coopers & Lybrand is known for its expertise in business advice. High-tech clients include Atmel Corp., Genus Inc., Cisco Systems Inc., Triad Systems and California Biotechnology Inc. Coopers & Lybrand would be an excellent choice for the engineering- or technology-based start-up entrepreneur. Contact Alan L. Earhart, Coopers & Lybrand, Ten Almaden Blvd., Suite 1600, San Jose, CA 95113. Phone (408) 295-1020. Additional offices: Boston, Paul Joubert (617) 574-5000; Los Angeles area, Melanie McCaffery (714) 251-7200; San Francisco, Cynthia Deldman (415) 957-3000.

Price Waterhouse

(Ask for *Taking Your Company Public* and *Expanding Into Exports.*) High-tech clients include Applied Materials, Hewlett-Packard, Biogen Inc., Connor Peripherals, and Borland International. Contact Benjamin Brussell, Price Waterhouse, 555 California Street, San Francisco, CA 94104. Phone (415) 393-8500. Additional offices: San Jose, Michael Patterson (408) 282-1200; Boston, Patrick M. Gray (617) 439-4390; San Diego, Thomas E. Darcy (619) 231-1200.

KPMG Peat Marwick

(Ask for *Business Planning* from KPMG's Private Business Advisory Services.) High-tech clients include National Semiconductor, Phillips, Motorola, Siemens, and Sequel. Contact Samuel J. Paisley, KPMG Peat Marwick, 1755 Embarcader Road, Palo Alto, CA 94303. Phone (415) 493-5005.

Ernst & Young

(Ask for the *Ernst & Young Business Plan Guide*, *Outline for a Business Plan*, or buy Daniel Garner and Robert Conway's related book, the *Ernst & Young Guide to Raising Capital.*) High-tech clients include Apple Computer, Sun Microsystems, Intel, Tandem Computers, and Genentech. Ernst & Young is the acknowledged leader in professional services for high-technology companies. Contact Roger Dunbar, Ernst & Young, 55 Almaden Blvd., San Jose, CA 95115. Phone (408) 947-5500. Additional offices: Palo Alto, Dave Ward (415) 496-1600; Palo Alto, Kenn Lee (415) 858-0505; San Francisco, Mark Pickup (415) 951-3331; Walnut Creek, Mark Pickup (510) 977-2907.

Arthur Anderson & Co.

(Ask for *An Entrepreneur's Guide to Starting a Business*, *The Life Cycle of a High Technology Company: A Guide for Success, An Entrepreneur's Guide to Developing a Business Plan*, *An Entrepreneur's Guide to Going Public, Compensation and Strategies for Corporate Directors, Effective Executive Compensation—A Competitive Advantage, Employee Stock Ownership Plans: An Executive Overview and Executive Compensation Strategies.*) Arthur Andersen is the largest accounting firm in the U.S. High-tech clients include Cadence Design Systems Inc., WYSE Technology Inc., Oracle Corp., Amdahl Corp., And Acuson Corp. For Silicon Valley residents, contact Mark Vorsatz, Managing Partner, Arthur Andersen, 333 West San Carlos Street, Sutie 1500, San Jose, CA 95110. Phone (408) 998-212. Additional offices: San Francisco, Thomas B. Kelly (415) 546-8200; Oakland, Marvin A. Friedman (510) 238-1320.

Deloitte & Touche

High-tech clients include Atarti, Microsoft, Syntex Corporation, Rockwell International, and 3Com. Deloitte & Touche specializes in providing services to high-technology growth companies. In Silicon Valley, contact Mark A. Evans, Deloitte & Touche, 60 South Market Street, Suite 800, San Jose, CA 95113. Phone (408) 998-4000. Additional offices: Boston, David Elsbree (617) 261-8000; Los Angeles, Alan Frank (213) 688-0800; Cost Mesa, John Moulton (714) 436-7100.

Grant Thornton

High-tech clients include Televideo Computer Systems, Western Microwave, Scorpion Technologies, DocuGraphix, and DSP Technology. Contact Gary J. Gemoll, Grant Thornton, 150 Almaden Blvd., Suite 600, San Jose, CA 95113. Phone (408) 275-9000. Additional offices: San Francisco, Gary J. Gemoll (415) 968-3900; Los Angeles, Richard A. Stewart (213) 627-1717; Boston, Sanford R. Edlein (617) 723-7900.

SMALL BUSINESS ADMINISTRATION ASSISTANCE PROGRAM

The Small Business Administration (SBA) sponsors three major assistance programs for entrepreneurs.

Service Corps of Retired Executives (SCORE)

SCORE is an outstanding organization consisting of more than 12,000 retired and active executives in over 735 chapters and offices across the country. These individuals, having successfully completed their own active business careers, are very willing to provide free advice to start-up entrepreneurs. Use them to your full advantage. SCORE is one of the best-kept secrets and most underutilized organizations around. Look up either SBA or SCORE in your phone book to find a contact.

Small Business Development Centers

SBDCs offer start-ups and growing companies a variety of free services. SBDCs individual state headquarters are usually located in the business school of a university. Sub-centers are located throughout each state. Some centers serve all clients in a region, and other offer specialized expertise to the whole state.

Small Business Institutes

SBIs serve only existing businesses and consist of teams of business school seniors along with graduate students and their professors. These teams will conduct a management audit and provide a confidential case report of your existing business for about $200, and might include a full marketing plan or a focus on one specific area of concern.

SMALL BUSINESS ADMINISTRATION PUBLICATIONS

SBA is an excellent source of more than 50 free and low-cost introductory pamphlets and manuals covering many aspects of starting a small business. Ask for the latest *Directory of Publications,* which is free from any SBA office (see your local phone book), or by calling (800) U-ASK-SBA, which would connect you to the

SBA's Office of Public Communications. Or, you could write to Small Business Administration, Office of Public Communications, 409 Third Street SW, Washington, DC 20416. While the SBA is geared toward the small business, its resources are well worth utilizing.

INCUBATORS

To assist start-ups, many cities offer incubation facilities, which consist of communal office space at reduced rates and a host of business services. To find the incubator closest to you, contact the National Business Incubation Association, 1 President Street, Athens, OH 45701. Phone (614) 593-4331, fax (614) 593-1996. Dinah Adkins is the executive director representing 620 members.

The Wall Street Journal (August 9, 1991) reported that the number of incubators has grown to about 450 from 50 in seven years, and they house an estimated 7,500 nascent companies.

CENTER FOR ENTREPRENEURIAL MANAGEMENT

Joseph R. Mancuso, who has written extensively on business, runs the Center for Entrepreneurial Management, a nonprofit organization consisting of about 3,000 members. Dues of $96 give you the opportunity to network with other entrepreneurs and also a subscription to *Success*. Write to Center for Entrepreneurial Management, 180 Varick Street, Penthouse, New York, NY 10014-4606. Phone (212) 633-0060.

AMERICAN MANAGEMENT ASSOCIATION

The AMA's Growing Companies Program is directed toward small businesses. Many of the AMA books, self-study guides, audio- and videotapes, seminars, and conferences are excellent, although not always cheap. Contact the American Management Association, 135 West 50th Street, New York, NY 10020. Phone (212) 586-8100. Individual membership fees are $160.

UNIVERSITY ASSISTANCE

A couple of university programs can help you get a jump start, regardless of your location.

Massachusetts Institute of Technology

MIT's Enterprise Forum Inc. will analyze product ideas, business plans, or entire companies for about $200. Contact Paul E.

Johnson, Director, MIT Enterprise Forum Inc., Massachusetts Institute of Technology, 201 Vassar Street, Room W59-219, Cambridge, MA 02139. Phone (617) 253-8240, fax (617) 258-7264. Also operated by the MIT Enterprise Forum is the Venture Capital Network. This group can play the role of matchmaker, linking wealthy individual investors or "angels" with entrepreneurs needing start-up cash. See Chapter 12 for discussion on angels and on obtaining funding for your start-up. A fee is charged to both investors and entrepreneurs for a subscription to the network. They also can provide the names of many affiliated organizations. For more information write to the Venture Capital Network, c/o the above address, or call (617) 253-7163.

James Madison University

The center for Entrepreneurship at James Madison University, and SBA Small Business Development Center, provides a national service called Innovation Evaluation Program. For $100 they will test the feasibility and patent potential of a new idea or product using a computer model. Contact Roger Ford, Director, Center for Entrepreneurship, James Madison University, College of Business Building, Harrisonburg, VA 22807. Phone (703) 568-3227.

Chapter 2
TECHNOLOGY-ORIENTED PROFESSIONAL AS COMPANY FOUNDER

"At a start-up, life suddenly seems unfair. A deadline is a deadline. Employees don't have several layers of management to buffer them from the outside world."
—T. J. Rodgers, 1986, Cypress Semiconductor

To achieve a successful start-up, portray yourself as your business' founder and chief executive officer. You will see a picture of excitement, opportunity, and challenge. Let us look more closely.

FOUNDER'S ROLES AND RESPONSIBILITIES

As founder of your own company, your business role initially spans the entire spectrum. Until you have the financial resources to add a competent staff in which you have great confidence, you are responsible for everything from managing the business and developing your ideas to selling the final product. Your duties will include everything from signing checks to emptying trash cans, and you must attend to all the administrative details of starting and maintaining a business. There are also legal issues in incorporating and licensing the business. Insurance, taxes, rents, utilities, bank accounts, and letterhead and business cards will all need to be taken care of during the first week. These things, however, can be fun to work on and might even be delegated to others.

Some much tougher problems represent your first real challenges—managing the five elements of a successful start-up:

1. creating your management team and board of directors
2. evaluating markets and targeting customers
3. defining and developing your product
4. writing your business plan
5. raising funds

Unless you start your company with a seasoned management team, you are the boss, and the burden of each of these elements falls on you. That is why you probably will give yourself the titles of *chief executive officer* and *president.*

OFFICES

You need to be familiar with the key offices of a corporation and with the roles and responsibilities of those holding such positions. In California, the secretary of state would require you to complete a statement by domestic stock corporations designating your company's CEO (i.e., president), secretary, and CFO (i.e., treasurer). The corporation must have these three offices in accordance with corporations code. Any [or all] of the offices may be held by the same person unless the articles [of incorporation] or bylaws provide otherwise.

Chairman of the Board

The *chairman of the board* is the member of the corporation's board of directors who presides over its meetings and who is the highest ranking officer in the corporation. The chairman may or may not have the most actual executive authority in a firm. In some corporations, the position of chairman is either a prestigious reward for a past president or an honorary position for a prominent person, a major stockholder, or a family member. It may carry little or no real power in terms of policy or operating decision-making. In a start-up, the position of chairman of the board is usually initially held by the founder, at least until a sophisticated investor helps fund the company and requests to occupy that position.

Chief Executive Officer

The title of *chief executive officer* (CEO) is reserved for the principal executive. The CEO is the officer of the firm who is principally responsible for the activities of the company. The CEO designation is usually an additional title held by the chairman of the board, the president, or another senior officer (such as a vice chairman or an executive vice president). In most seed-stage and start-up companies, the founder is both the president and the CEO. (Chapter 6 covers the distinctions between seed, start-up, and the various other stages of new companies.)

President

After the chairman of the board, the *president* is the highest-ranking officer in a corporation (unless the president also holds the CEO title, in which case the president and CEO can have more actual executive authority than the chairman). The president is appointed by the board of directors and usually reports directly to the board. In smaller companies the president is usually the CEO, and exercises authority over all other officers in matters of day-to-day management and policy decision-making.

Chief Operating Officer

In larger corporations the CEO title is frequently held by the chairman of the board. This leaves the president or an executive vice president as the *chief operating officer* (COO), responsible for personnel and administration on a daily basis. The COO reports to the CEO and may or may not be on the board of directors. (COOs who are presidents typically serve as board members, while COOs who are executive vice presidents usually do not.) The COO title is also often used in recognition that an operations person has taken on increased responsibility from the president or CEO.

Chief Technical Officer

The title of *chief technical officer* (CTO) is a curious one. It is widely used in Silicon Valley to recognize key individuals upon whom a company is clearly dependent for technical contributions. If you are an engineer with an idea for a product and want to exploit that product in a business start-up, you do not necessarily have to be the CEO and president. Do you really want to manage? If not, maybe you would rather be the CTO. The remainder of this book explores the ramifications of taking on operations management responsibility and your preference should be clearly established by the time you get to the last chapter.

Vice President of Engineering/Research and Development

The position of vice president of engineering is not one to be underestimated. It involves a substantial challenge, and your company's success will depend upon this person's ability to deliver your product on specification, on budget, and on time.

For the vice president of research and development, schedule pressures are not as severe. This person is responsible for developing the technology needed for future generations of your company's products. Most start-ups cannot afford both a vice president of engineering and a vice president of research and development, in which case one person must serve both roles. The company's long-term future rests on this person, and whoever holds this position should be confident in his or her ability to produce future products.

Would you be happy starting your own company, but not holding the top slot? This is a very important question for you to consider. If you want to be successful but have little management experience, it makes a lot of sense to start up with other experienced management. Together, you may all make a lot more money than any one person could alone. We will talk more about management teams later, but plant this option in your mind. Who do you want to be in your new company; what role do you want to play? What will make you happy, while allowing you to meet your financial objectives?

FOUNDER CAREER PATHS

The fact that you are reading this book implies that starting your own business is obviously an idea at the top of your list. However, while you might be determined to start your own business and hold the positions of president and CEO, you may also have a preplanned career path that eventually puts you back into a technical position where you would be most comfortable and happy. This might also make other key employees and your investors quite happy. Investors might even insist that you make such a transition as a condition for continuing to fund the business, especially if they are later unhappy with your performance. One perfectly reasonable career path would lead you initially to start and run your company as president and CEO, and later to assume the position of vice president of engineering or vice president of research and development.

A typical scenario for an engineer founding a high-growth, technology-based company involves first launching that company by holding the president and CEO titles. Later, bringing in professional management and funding may provide the opportunity to move into the vice president of engineering or vice president of research and development role, where one might be most productive and comfortable. The title of chief technical officer would be appropriate in such a situation. Many founding engineers get forced back into those roles either as a condition for obtaining initial or additional venture capital funding, or as a result of poor operating performance. If you want to be the CEO and president, and are up to the task, by all means go for it. However, if you lack the experience or disposition, why not plan to hold a position you would really enjoy?

Michael S. Malone's book, *Going Public*, which describes the successful Initial Public Offering (IPO) of MIPS Computer Systems, Inc., quotes the founders as saying,

> One of the things we did right was to recognize that we weren't going to be management. We could help in technology or any place

> else we were needed, but we were not businessmen. We had seen the stories of big egos that invent something wonderful and then think they can be CEO—and they can't, and it crashes the company. None of us aspired to run the company and that's why we've got this marvelous management team in place.

Learn from others' successes!

ENTREPRENEUR'S PROFILE

Edward B. Roberts' recent book, *Entrepreneurs in High Technology*, contains the most comprehensive study to date of personality traits of the engineer creating a start-up. Many of the characteristics listed below are loosely based on Roberts' scientific observations. If you can associate with many of these statements, then you know this book was written for you, as these are the characteristic influences on becoming a technical entrepreneur.

- I have a long-felt, strong desire to start my own business.
- I had a self-employed father.
- I have a minimum of a four-year undergraduate degree, or a master of science degree in some technical field.
- I think I can do a better job than others in delivering a service or in producing a product. I am willing to work hard for something that is important to me.
- I am very independent and have a continuing need to meet and overcome challenges.
- I have only moderate needs for group achievement and power, and a low need for affiliation. (There is a psychological theory that says everyone holds two of three needs more closely than the third: achievement, power, or affiliation. For example, if you value affiliation with associates and achievement of group results, it is unlikely that you can lead from a position of real power. As an analogy, you cannot maximize both revenue and profits—something has to give.)
- I have at least a decade of work experience.
- I have published more papers and obtained more patents than my associates. I am highly productive.
- In my work, I carry out applied development work, not research.
- In my career, I have already risen to managerial levels.
- I have only modest concern for financial rewards (at the time of start-up).

- I am more extroverted than my technical associates at work, but relative to the rest of the world, I still look more like an introverted inventor than a businessperson.
- In my current place of work, I feel challenged and find satisfaction.
- Although I am an engineer, I tend to buy more books on business than on technology.
- I read the business section of the newspaper nearly every day.

Having these characteristics does not necessarily mean you will be a successful entrepreneur. A strong orientation toward marketing is necessary and is emphasized throughout this book.

WHAT IS NEXT?

The following chapters further describe life in your start-up, and emphasize the importance of teams, high-growth, and a customer- and market-focused versus technology-focused strategy. Studying unfamiliar business and financing terms and visualizing approaches to solving start-up problems will prepare you to create a management team, identify customers and markets, define products and services, write business plans, and obtain funding. Essential wealth-building tools and strategies are extensively described, educating you about unfamiliar yet essential practices involving stock grants, stock options, and other instruments.

Chapter 3
LIFE IN YOUR START-UP

"Few people do business well who do nothing else."
—The Earl of Chesterfield

In this chapter we present some statistics and observations that candidly lay the facts on the line: what are your chances of being successful and happy in a start-up? Some individuals thrive in the excitement of the fast lane. Others encounter an inability to cope with the overload, stress, and constant change start-ups can entail.

SUCCESS AND FAILURE: STATISTICS

What will happen if you launch a start-up—how will your life differ? First, do not assume you will even be able to get started. Optimistically, perhaps only 10—30% of start-ups that seriously look for venture capital actually get funded. Realistically, the number might even be much lower. Any one venture capital firm will typically fund only a fraction (0.6%) of the hundreds or thousands of business plans it receives in any given year. Even should you get the business started, do not take personal success for granted.

- It has been estimated that less than half of the entrepreneurs who start companies surviving five years or more actually remain with their start-up.
- If your company is funded, becomes successful, and goes public, you will earn about $6.5 million within five years.
- Only 10% of venture capital-funded start-ups go public.
- Sixty percent of venture capital-funded high-tech companies go bank rupt.
- Founders of typical high-tech companies own less than 4% of the company after the initial public offering, far less than 10 years ago.

According to Drew Field's book *Take Your Company Public*, in the past few years there have been roughly 700,000 new incorporations each year.

However, there have been fewer than 300 initial public offerings—that is only a 0.04% success rate for going public. Keep in mind, however, that this is a distorted figure since it includes all sorts of small businesses.

Michael S. Malone's book, *Going Public*, states that in Silicon Valley, "of the ten thousand or more companies that have been founded in the last three decades, no more than a hundred have gone public." That translates into about a 1% rate for engineering-related start-ups that also successfully complete an IPO.

Edward B. Roberts' *Entrepreneurs in High Technology* also provides a great deal of valuable data.

- The actual failure rate of high-technology companies founded by MIT associates is only 15—30% over the first five years.
- Having a Ph.D. degree leads to more failures than for those with master's degrees, except in certain fields such as biotechnology.
- It may be harder to completely fail than imagined. Only about 20% of the large number of MIT spin-off firms ever are liquidated or go bankrupt. When the press mentions that about 80% of all businesses fail, remember that many of these are small businesses such as gas stations and used car lots.

The so-called living dead make up the majority of the surviving start-up endeavors. These companies do not fail in that they go bankrupt or go out of business, but they do not really succeed either. While some may provide an interesting and stable living for their employees, many others provide low salaries, no capital gains on the company's stock, no retirement funds, and no vacations. These employees might have been better off staying with their former companies. It is important to know when to bail out of a living dead situation.

VACATION AND TIME OFF

More than likely, you will be totally consumed by your start-up business for the first few years. It will be with you every hour of every day. It will seem at times that there is no end to it. The old saying "it ain't over 'til it's over" aptly applies. Any scheduled break, weekly recreation time, or short vacation will provide only a brief respite from the pressures of your business. Bill Gates, founder of Microsoft Corporation, took only two three-day vacations during the first years of his start-up. Even later, he took one week of vacation a year. Do not plan on taking two- or three-week vacations during your first few start-up years either.

The 1991 MasterCard BusinessCard Small Business Survey study of small businesses showed that 22% of their owners took no vacation, as detailed in Figure 3.1.

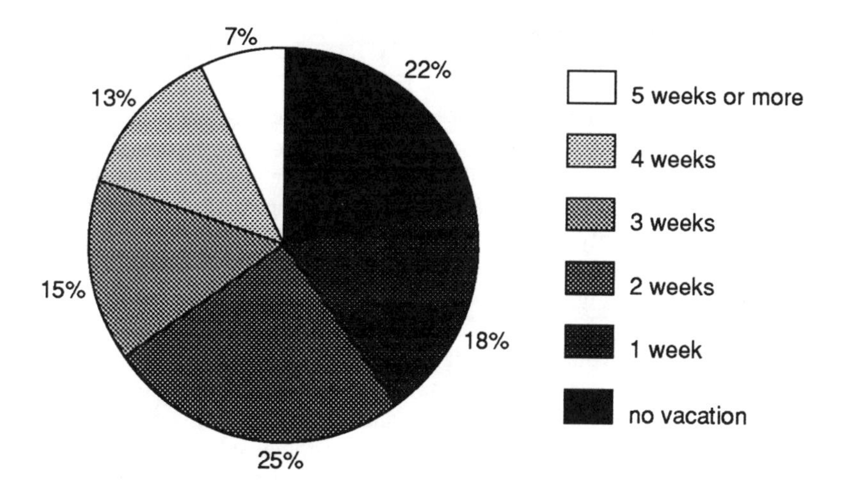

Figure 3.1 Annual Time Off for Small-Business Owners

A *small business* was defined here as a firm with at least one employee but less than 100 employees. According to the U.S. Small Business Administration (1988), there are slightly fewer than 4 million such firms in the U.S.

It is well known that small-business owners find vacations hard to take. They fear customers will evaporate. They are reluctant to trust someone else to fill in, and they will convince themselves they are indispensable.

As a start-up entrepreneur you will feel many of these same pressures. You have only so much money and time to complete your product, produce sales, and generate revenues. Will you take two weeks off to recharge during your concept development, seed, or start-up phases? Probably not. You may very well go three to ten years without taking a real vacation. Even if you do get a few days off now and then, you will likely find that you need or want to be in constant communication with those in the office.

It is difficult to imagine the CEO today who does not have a cellular phone in his or her car, and who is not in constant touch with the organization, its customers, and its investors. Home facsimile machines and portable computers with modems for accessing electronic mail (ubiquitous today) truly make getting away from work impossible. The fact that you will probably take your computer with you on a family vacation could also have an impact on your family. They may wonder why you even came along if you have to work so hard.

However, it is possible to remain very calm and relaxed as you start a venture, acting on the notion that it is good for the business to get away and recharge yourself at regular intervals. In this way you may do well enough, but you most likely will not maximize the growth and valuation of your business. Putting recreation ahead of business is a quality of life issue, a priority only you can establish. Just be aware that there will be many internal pressures within yourself, along with external pressures that you cannot always ignore.

For example, are you really going to skip that next board meeting where you need to ask for some more funding? Is it a good idea to turn down the invitation to attend your lead VC's annual CEO meeting? Should you send someone else to talk to that critical customer who is considering taking his business elsewhere? As a start-up entrepreneur, you are an integral part of your new company, and you will find that you need to be available to handle such situations.

Perhaps if you have one or two million dollars in the bank, are still on plan, and have no exceptional problems in the business, you might be able to go skiing over a long weekend or head off to Club Med for a week. More likely, though, you will be coming up with yet another new business plan and looking for more money, cultivating customers, finding employees, lining up distributors, setting up manufacturing or service operations, conducting a press tour, or preparing for a trade show. It will seem that there is never enough time. But for a dedicated, intelligent, enthusiastic person, the stressful period in a start-up is greatly outweighed by the sense of accomplishment and fulfillment of creating your own successful business.

WORKING HOURS

A 1991 *Inc.* magazine article showed that 66% of the founder CEOs of the fastest growing *Inc.* 500 companies worked at least 70 hours a week while the company was getting started. Only 13% were still doing so five years later. Figure 3.2 details the results of the *Inc.* 500 study. The only founders who managed 40 hour or less workweeks were presumed to have been involved in other businesses at the time of start-up.

As one drives past the many high-technology business parks housing start-up companies in Silicon Valley, it is not unusual to see cars in the parking lots on Saturdays and Sundays. Workweeks of six or seven days are more common than the standard five-day workweek.

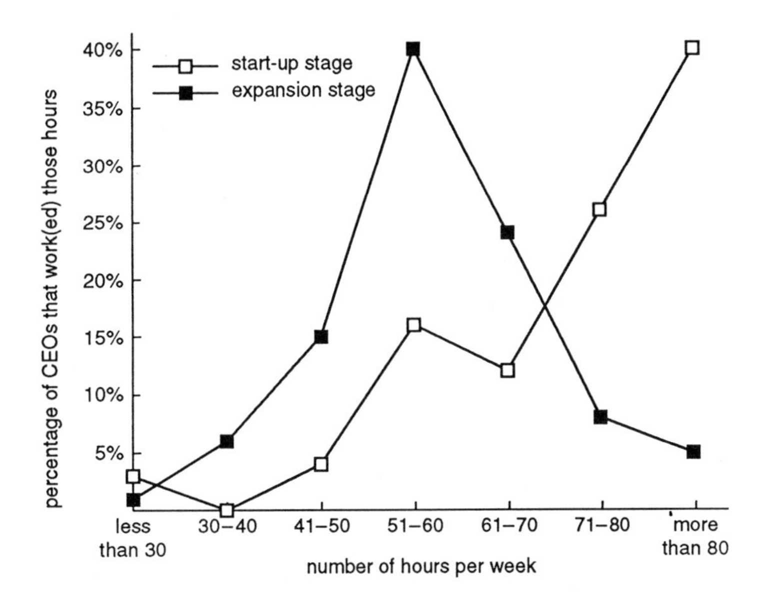

Figure 3.2 Anticipated Weekly Time Commitments of CEOs

Evidence suggests that start-up entrepreneurs typically work well over 10 to 12 hours a day, six to seven days per week. That is almost double the number of hours put in by the typical *Fortune* 500 employee. In fact, in your start-up, you might work even more than this, considering that every night you will bring something home. If nothing else, you will have problems to ponder, or work-related reading. Gates of Microsoft, for example, found himself regularly working 65 hours a week, even after his company went public. More than once start-up entrepreneurs in the hard-driving, high-technology Silicon Valley may find themselves working around the clock to meet an important deadline. It is not unusual to find sleeping bags tucked away in a cabinet for those nights when it is too late and you are too tired to make it home safely, especially when you have to be up and at it in a few hours anyway.

If you already work long hours like this for someone else, then you will be in good shape for starting your own company. Working hard can be rewarding, especially if it is for your own business.

DIVORCE

Start-up entrepreneurs have a high divorce rate. They may prefer being with their businesses more than their spouses. The pressure and stress of a start-up can tarnish any relationship with time. This is not to say, however, that start-ups cause divorces. It is more likely that start-up entrepreneurs typically have more reasons for divorce. Roberts' observations reinforce this notion, but he also found empirical data hard to come by. You should, however, think seriously about what impact starting your business will have on those around you.

It is essential to have the support and encouragement of your spouse during a start-up. Sharing your vision and mission with a supportive loved one is going to be very important to you.

HOLDING YOUR BUSINESS TOGETHER

William H. Davidow, a famous Silicon Valley venture capitalist and author, states, "There has to be something that holds a company together beyond making money...the glue is that your people better love what they're doing, better be committed to the mission, [and] better believe in winning." While Davidow was addressing the problem of managing MIPS Computer Systems, Inc., after going public, his words apply to every phase of your start-up. This is your leadership challenge: to impart this same love, commitment, and belief to your team! Chapter 8 will discuss further the importance of building such a team, and how to make it happen.

PERSONAL PLANNING PROCESS

When considering whether (and if so, how) to continue with your venture, you will want to examine the personal side of starting your own business. You need satisfying answers to questions that might not come out during the traditional business plan writing process described in Chapter 11.

First, you should clearly establish an alignment of the compelling interests of yourself and your founding team. What are the personal driving forces behind your wanting to start this venture? Do you each share a common vision of the company's mission? What product will be built? Who will be the customers? How fast and how large will the business grow? Apple Computer, for example, set out to create an insanely great product. All the founders believed that, and their success followed. However, you need to do more than simply recognize the need for establishing this common vision and alignment of interests. The remainder of this book should guide you toward achieving that goal.

Second, you need to objectively assess the motivation and expectations of the founding team.

Do not jump into the business plan writing mode until you have thought out the preceding questions.

ALLOCATION OF EFFORT

Roberts measured the allocation of effort in technology start-ups by engineer-founders during their first six months of business. He found that less than one-third of their time at work was spent in engineering, as shown in Figure 3.3.

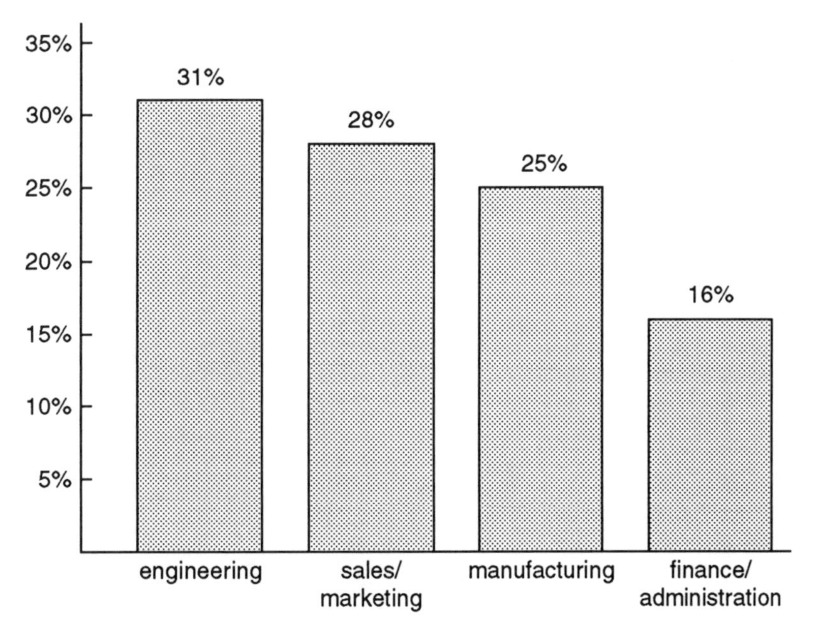

Figure 3.3 Effort Allocated by Founders During First Six Months

The lesson here is that, while you may be successful, you may not always spend your time doing what you thought you would be doing. The following chapters discuss how you can make your business successful, rewarding, and fun.

Part Two

GETTING DOWN TO BUSINESS

Part Two of this book deals with some important concepts that can make or break your start-up endeavors. As an engineer, you especially must develop a keen appreciation for the essential role of marketing as a focus for growing your business. You must realize that growth will be essential for your success, and you should learn the basic finance-related terminology you will encounter as you launch your start-up.

Chapter 4
MARKET-VERSUS TECHNOLOGY-FOCUSED APPROACH TO GROWING A BUSINESS

"Marketing is essentially viewing the enterprise from the viewpoint of the customer, and there is very little difference between it and the management of the enterprise as a whole."
—Peter F. Drucker

BENEFITS TO CUSTOMERS

If your technology enables you to quickly develop a unique product that customers will purchase, satisfying a vital market need and leading to rapid profitability, then you are on the right track. However, engineers are especially apt to neglect the essence of a successful business, which is delivering benefits to customers. Delivering benefits is not the same as selling technology.

As Figure 4.1 depicts, the elements of a successful business include, among other concerns, technology and markets.

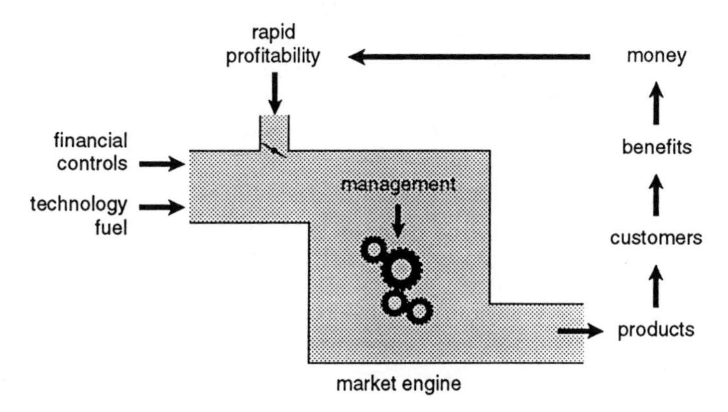

Figure 4.1 Market- and Customer-Driven Technology-Fueled Business Machine

TECHNOLOGY AND MARKETS

Technology is certainly an essential element of an engineering-related start-up. However, to focus your business on your technology alone is foolhardy. Understanding and exploiting markets is also an essential element of any start-up.

Market Positioning

Where does your proposed product fall in Table 4-1?

Table 4-1 Market Positioning

	existing product		new product	
new market	**II**	**marketing driven (new use for an existing product)**	**I**	**missionary sales, technology push**
existing market	**III**	**face entrenched competition**	**IV**	**market-driven, technology-fueled, market pull**

If your product falls in any quadrant other than quadrant IV you need to be careful. An examination of each quadrant follows.

In quadrant III, the *existing product/existing market* quadrant you would be a new entrant in the marketplace. Although perhaps acceptable for an income substitution business, this is probably not a lucrative market position for a technology-oriented engineering start-up. Opening a new restaurant would fit into this category.

Quadrant II, the *existing product/new market* quadrant, represents an opportunity for a business with superior marketing and selling skills to create new demand for an existing product. Typically, engineering-related firms will not play in this position. An example is 3M, which found lucrative markets for tape-based products. 3M's Post-itTM brand self-stick removable notepad is an excellent example of marketing an existing technology (glue and paper) to a new market (almost everyone who works in an office).

Quadrant I, the *new product/new market* territory, is the classic yet very difficult path taken by pioneering, technology-driven entrepreneurs. These individuals take all the arrows in their backs, often only to have new market entrants quickly exploit their expensive ground-breaking efforts. Of course, there have been some big successes in this area, but there have also been some very big failures.

The first video games and home computers are examples of new products that played in new markets. The markets in these cases absorbed almost everything offered. However, early players often had a rough beginning. Texas Instruments, for instance, offered a home computer that was wonderful for playing games. But it had to be patient and dig deep into its pockets because it took a couple of years to get the FCC approval that finally made the new products attractive to the market. Other ventures (such as Trilogy's attempt to create wafer-scale integration electronics for computers) consumed hundreds of millions of investors' dollars before ultimately failing.

Quadrant IV, the *new product/existing market* quadrant, is the safest category. Here, you can leverage your engineering technology to produce an advanced, new product, delivering more benefits at lower cost to customers in a market that is not only receptive to but demanding your new development. Here you let the market pull you into deciding which product to develop. Do not push your technology onto a resisting market.

Technology Push

A business focus based on technology is often called a *technology push* because you are relying on your new technology to push customers into a new market.

Engineers, especially, are likely to have ideas for products never before imagined. Our discussion throughout this book on the importance of identifying customers and markets for your products emphasizes your need to have a commanding position in a protected market niche. You do not need a new product to do that. Instead, you can simply "do the common thing uncommonly well," says Paul Oreffice, chairman of Dow Chemical. A new concept is usually evident to the market, and it is difficult to develop any fresh idea in total secrecy. Also, unless you have an enforceable patent, you would not have a significant advantage over rivals entering into competition with you. A really new idea will place you in technology-push territory where you will be forced to do missionary sales, which require educating the market. This delays the profitability of your idea.

A good example of the difficulty of pushing brand-new technology into new markets to achieve business success is reflected in the experiences of Xerox Corporation. In many of Xerox's attempts to exploit new markets using the most advanced technology, other companies ultimately reaped the rewards when the markets were ready.

Market Pull

Many high-technology start-ups are finding that instead of pushing breakthrough hardware technology onto the market, they are better off by participating at the commodity level and distinguishing themselves through exceptional service and cost. Niche-oriented markets with ever-shortening product life cycles characterize the 1990s. Products must increasingly be designed for world markets. From inception, growth businesses will need to compete globally.

A viable business model is to recognize unique market opportunities that slightly stretch state-of-the-art technology, and then develop products based upon these identified market needs and your technological capability.

A business focus based on market need is often called *market pull* because you are relying on the market's desire for a specific benefit to be met through an expansion of state-of-the-art technology. In market pull, the customers are ready and you must deliver the technology. In technology push, on the other hand, the presence of customers is questionable even if you can deliver the technology.

Market- and Customer-Driven Technology-Fueled Strategy

The key to launching a successful technical start-up is to make maximum use of any proprietary technology you can develop while, at the same time, focusing on market opportunities. This is called a *market- and customer-driven technology-fueled* strategy. The term *market-driven technology-fueled* was first embraced and popularized by Measurex Corporation in the 1980s to guide its business.

New Markets

If, instead, you have a new technology focus, you may be forced to create new markets to sell your products. Creating a new market for a new product is a very difficult and expensive task. It will take extra money and extra time to reach a break-even point if you are creating both a new product and a new market. You must strive for rapid profitability.

Risk and Reward

Of course, the less market risk you entertain, the fewer rewards you might expect. For example, if in the early 1990s you chose to compete in the personal computer mass storage arena, you had several choices of products you could develop. As depicted in Figure 4.2, a low-risk, very low-reward strategy would point you toward an existing "me-too" stable product such as the 3.5-inch hard disk drive market. Clearly this market lags behind technology capability. A more moderate avenue would have you developing 2.5-inch hard disk drives for notebook and pen-based computers, but you would have lots of competition and the rewards would be tempered. A higher risk path with higher potential rewards would have you developing the smaller sub-two-inch drives, already in development, and clearly desired (pulled) by the market for even more portable and compact computers. Slightly more risky would be a pure market pull memory card substitute for rotating media storage devices. On the other hand, a technology push product approach would have you developing exotic memory substitutes like holographic memory modules.

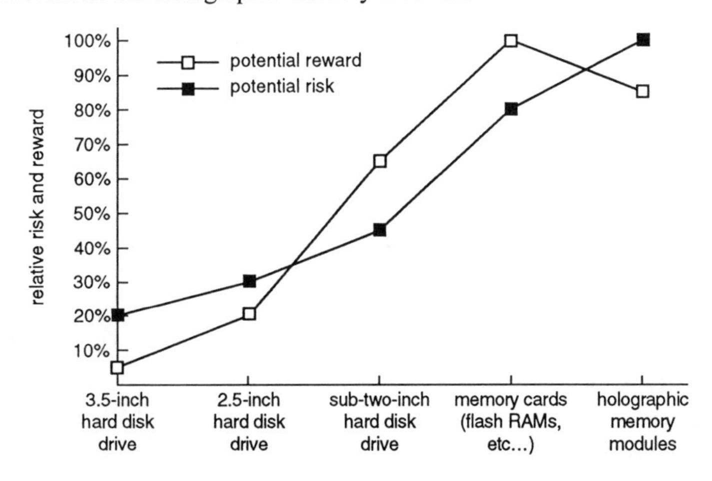

Figure 4.2 Subjective Plot of Potential Risk and Reward

After you have itemized your alternatives, make sure that:

1. your potential rewards exceed your risks
2. you maximize the spread between potential risk and reward
3. the market does not lag your technology capability
4. your technology is pulled by the market
5. you do not push your technology on the market

In our example, the technology-focused holographic memory module approach might be most attractive to you as an engineer, but the current market clearly desires more down-to-earth products such as memory cards and smaller disk drives. Deal with your technology-related product ambitions in manageable risk stages. If you take on a risky product, make certain that you have the financial backing to complete it. The greatest mistake you can make is to start a business that you cannot finance. If you need to raise money to move on to the next stage of risk, make sure that you secure adequate financing. If you run out of money, you will lose all control of your business.

RAPID TIME TO MARKET

Rapid time to market and rapid profitability are commonly recognized as keys to start-up success. Time consumption, like cost, is quantifiable and therefore manageable. Today's new-generation companies recognize time as the fourth dimension of competitiveness and, as a result, develop new products rapidly, operate with flexible manufacturing and rapid-response systems, and place extraordinary emphasis on research, development, and innovation. Organizations are structured to produce quick responses rather than low costs and control. Figure 4.3 illustrates the mathematics of the time-value of money with a conceptual view of the different capital investments that would be required to achieve break-even in the years indicated.

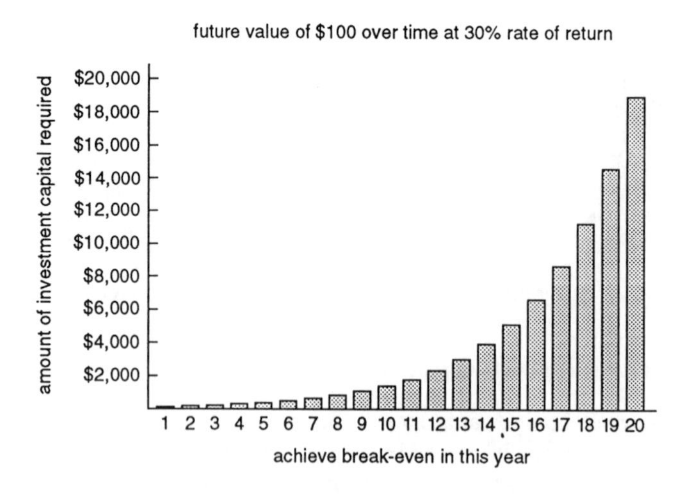

Figure 4.3 The Time-Value of Money

The figure depicts a typical venture capitalist desiring returns of 30% compounded annually. You can see that if you double your time to break even, you will exponentially increase the investment needed.

By becoming a time-based competitor, your start-up can get there first and develop a distinct competitive advantage over more technically sophisticated companies that attempt to push the market.

Symantec Corporation: Responding to a Changing Market Need

In the early 1980s, a large number of high-technology companies were launched in the area of artificial intelligence (AI). AI included robotics, computer vision, and natural language understanding. Each of these areas drew substantial investments but almost none returned any rewards. It was a classic case of technology push. The market in the 1980s for products derived from AI technologies was virtually nonexistent despite the millions of dollars poured into it.

Symantec Corporation originally intended to develop AI products that enabled a computer to understand natural language. Potential applications ranged from natural language translators for the military and tourists to typewriters to which one could speak. Unlike most AI start-ups, Symantec's new management finally realized that the market was not responding to its technology and shifted focus before it was too late. Today, Symantec is a leading public company in the personal computer utility software market. (You may recognize it not only for it well-known SAM software to ward off computer viruses, but also it acquisition of Norton Utilities.) Its new market has proven to be one of the highest margin niches venture capitalists could have invested in during the early 1990s.

Symantec was successful because is responded to market need. This is an interesting case because Symantec seemingly abandoned its technology base. On closer examination, however, it was obvious that its marketable technology consisted of exceptional state-of-the-art computer science and programming skills that could be applied to pressing market needs. Added to Symantec's technology fuel was its savvy market-driven strategy, which resulted in admirable success.

Chapter 5
WHEN HIGH-GROWTH BUSINESS IS DESIRABLE AND NECESSARY

"A growing company in a growing market can survive a lot more management blunders and bad luck than a company in a stable market and still succeed."
—Gordon B. Baty

WHY GROW?

Understanding the importance of growth in business is something one learns. Few engineers have had the required training in finance and economics to internalize the need for growth, and the need is not obvious through introspection. Thus, one of the most difficult concepts for entrepreneurs to understand is why their new businesses must grow. It follows that one of the most difficult conclusions for an engineer to accept is that growing a successful business involves much more than engineering and technology. Often the CEO of an engineering-related start-up will be engaged in almost everything except engineering and product development.

The primary purpose of this chapter is to explain why growth is not only desirable, but necessary, if you are to achieve even modest financial objectives.

THE SELF-EMPLOYED

If you are a self-employed engineer and want to create wealth, you must focus on growth. Many engineers become self-employed (consultants) and motivated by losing a job. If you find yourself in this position, you need to understand the limits of your success as a one-person business.

The 1990s revealed an increasing number of self-employed entrepreneurial individuals. About 8.97 million people (7.7% of all workers) were self-employed. This was the highest level of self-employment in 25 years. Few of these people quit their jobs voluntarily. You will want to do better than these individuals; the average income from a business owned by one person was only $12,352 when doctors and lawyers were excluded. As you launch your technology-based start-up, whether motivated by choice or by losing a job, you will need to look to growth for financial success. Since one-person businesses (which, by definition, cannot grow) do not yield significant incomes nor do they create vast wealth, the successful entrepreneur must look for growth opportunities.

GROW A COMMANDING POSITION IN A DEFENSIBLE MARKET SEGMENT

According to William H. Davidow's *Marketing High Technology*, "Marketing must invent complete products and drive them to commanding positions in defensible market segments."

What this means in terms of growth is that your business must expand at least to the point where it can survive. A General Electric study showed that companies with a market share greater than 30% were almost always profitable, whereas companies with a market share of less than 15% almost always lost money. Since mathematically no more than six companies can have a greater than 15% market share, you must grow your business to one of a few commanding a 15—30% minimum market share. This does not mean you have to be a giant company to play in a giant market. Rather, you must identify segments of your market that you can dominate (that requires growth). Before you try to compete with General Motors or IBM, make sure you can garner a 15% minimum market share in a small, related, and well-protected market segment.

Joe Christenson, president of Pattern Processing Technologies, Inc., stated the case well for his company in a recent annual report.

> Despite a trend towards continuing consolidation in the late 1980s, the market is still fragmented with over 40 companies capturing a portion of the market. Our goal at PPT in the 1990s is to grow faster than the market and thus continue to increase our market share and successfully ride the trend toward consolidation.

Christenson clearly understands that growth is essential not only for extraordinary success, but also for mere survival.

ATTRACT CUSTOMERS IN EXPANDING MARKETS

Although it is understood that customers want to be sure their vendors will still be around in a few years to provide them with parts, service, software upgrades, etc., it is frustrating for the start-up engineer to make a sales presentation and then hear the customer say, "The product is just right, but will you be there next year to service it?" After all, how are you supposed to launch your new business if customers want to purchase only from proven sources? This is one reason you need to identify an unfilled market need, and to provide a product with several times the performance-price advantages over competitive products.

Many start-up businesses find good customers in *Fortune* 500 companies. Individual customers in these companies often associate with a struggling entrepreneur, and will help you over many hurdles. In return you need to help these individuals do their jobs well by providing a product that delivers on its promises.

DEVELOP A PRODUCT FAMILY

A single product does not constitute a business. Therefore, you must strive to provide a family of products to meet market needs. You should grow to provide solutions to different customers with differing needs. For example, a successful software company will need to make its software available on a variety of hardware platforms and operating systems.

Because customers are most often drawn to brand names, your product's trade name will only gain national or international recognition with volume sales and extensive publicity. Marketing communications and public relations efforts need to be amortized over a large, growing product base.

Since your first product serves a finite part of the market, you must grow to reach new markets with new needs. If you have only that first product, once your market is saturated your business ceases to grow.

ACHIEVE CRITICAL MASS AND ECONOMY OF SCALE

All the supporting infrastructure of your business, including your plant and equipment, management salaries, inventory control systems, accounting systems, and all other fixed costs, need to be spread out over your product base. Without a growing product base, the single-product enterprise will soon be overburdened with fixed costs. Until your company revenues exceed some critical mass, your fixed costs will restrain profitability.

Growth also permits you to add staff with specialized experience to your business. Without a sufficient volume of activity, you and your managers will be forced to become jacks-of-all-trades: you may be doing many things, but no one thing is being done well or efficiently. Since your business must be efficient to be competitive and profitable, it follows that you will need an expanded staff including various divisions of expertise. To attract top employees to your business you must provide both opportunity and financial reward. These, too, come from growth.

Economy of scale applies to every successful business. Henry Ford was perhaps the most famous entrepreneur to understand and apply the concepts of economy of scale to achieve critical mass, as he introduced mass production and the moving production line, transforming the early twentieth century handcrafted automobile business into a modern industry.

The Boston Consulting Group's 1968 book *Perspectives on Experience* argued that the cost of doing business decreased 20—30% every time business (sales) doubled. Although the BCG recommended that "if market dominance cannot be achieved, then an orderly withdrawal from the business is best," you will not always have to completely dominate a market. Many markets can support a number of successful companies, particularly if the markets themselves are growing.

DIVERSIFY TO DIMINISH BUSINESS RISKS

Though it has been said that you can always win the game of business as long as you do not place all your bets on one product, one customer, one supplier, or one investor, some attempts are going to fail. By diversifying your business, however, by producing a family of products and selling to customers in different industries, you will increase your chances for success. It is wise, though, to make sure you have established yourself on solid ground (i.e., established a 30% market share) before you start diversifying.

Your company must grow in order to diversify. This is not to say you should not stick to what you know best, however. As your business grows and reaches customers with different economic buying cycles and different problems to be solved, you will decrease the risk that a single catastrophic event could end your business.

Start-ups are exciting and risky, and you can thrive on that precariousness to make the most of your personal, financial, and emotional investment, but you do not want to persist in that state of uncertainty forever. It may be an oversimplification, but most start-ups are destined to either die, join the the living dead (i.e., barely subsist), or grow to success. For example, if you stay small and maintain a one-customer, one-product kind of business, you will someday have a cash-flow crisis that will bankrupt your venture. If you grow just to the size where your business breaks even, you have created an income substitution business. Finally, if you grow and never stop, you can achieve great wealth and security.

If you have to be in a start-up to get your kicks, do not make it the same start-up for a decade! Grow to success, and then restart if you have to.

CREATE CAREER OPPORTUNITIES

Great companies are run by great people. Great people are attracted to great companies. Getting, and then keeping, key employees is essential. Exceptional people need to grow, to learn, and to take on increasing responsibility. Only a growing company can provide such an attraction for the people you need. Almost everyone wants to be promoted to a better position, and those positions can only be created through growth.

CREATE FUTURE START-UP OPPORTUNITIES

One of the best sources of a start-up idea is one's previous employer. If your company grows, it will create an abundance of new ideas. Not all of these ideas can or should be exploited by the core business, and many employees will someday leave to create new start-ups, based in part on exploiting these untouched opportunities. These start-up opportunities can be yours if you choose. Or, your company can beneficially invest in those ideas that need a start-up environment to flourish by providing initial funding in exchange for equity ownership (in new start-ups).

CREATE MARKET VALUE, ATTRACT INVESTMENTS, AND CASH OUT

You and your investors will someday want to exchange your stock certificates for cash. To do this, your company must be of sufficient size and profitability to either go public or be acquired by another (usually public) company. In either case the result is that you and your investors would then hold marketable securities. This is normal and desirable—and it requires that your company grow.

Chapter 6
START-UP FINANCING TERMINOLOGY AND STAGES

"Money is the seed of money, and the first guinea is sometimes more difficult to acquire than the second million."
—Jean Jacques Rousseau, *A Discourse on Political Economy*

This book is concerned with the formation of your new company, which falls into the category of early stage financing—what is loosely called a *start-up*. Starting and developing your own successful high-growth company will take you through several major financing stages, known in investment circles as early-stage financing, expansion financing, and IPO/acquisition/buyout financing.

Each of these stages consists of key financing events. For example, in early-stage financing, there is a *seed* financing event that occurs when you first obtain funding to launch the business (or even just to explore a product idea or for research and development, long before there is a business), *start-up* financing, which is used for product development and initial marketing, and *first-stage* or *early-development* financing, which allows you to initiate manufacturing and sales. Many venture capitalists speak of seed financing events as comprising a separate investment stage because of the diversity and range of scope in these incubation deals.

The best way to characterize the stages of your company's growth is to speak the language of investors. Stanley E. Pratt's *Guide to Venture Capital Sources* and James L. Plummer's *Q.E.D. Report on Venture Capital Financial Analysis* converge on the investment community's various definitions. You must understand what these investment and company growth terms mean if you are to properly represent your situation to prospective investors, team members, and employees. Figure 6-1 is derived from Pratt's and Plummer's prose.

From Figure 6-1, it is evident that a start-up is a private company in a very early stage of maturation. It is thus interesting to note that the term start-up is often inappropriately used in reference to young public high-technology companies. Someone once said start-up is a state of mind. Perhaps some of these companies are able to retain the flair of start-ups, but most likely they would not have the risk, financial leverage, and reward opportunity, or the lack of infrastructure, instability, and uncertainty associated with a true start-up. Nor do they meet the start-up financing level description of "not having sold product" yet.

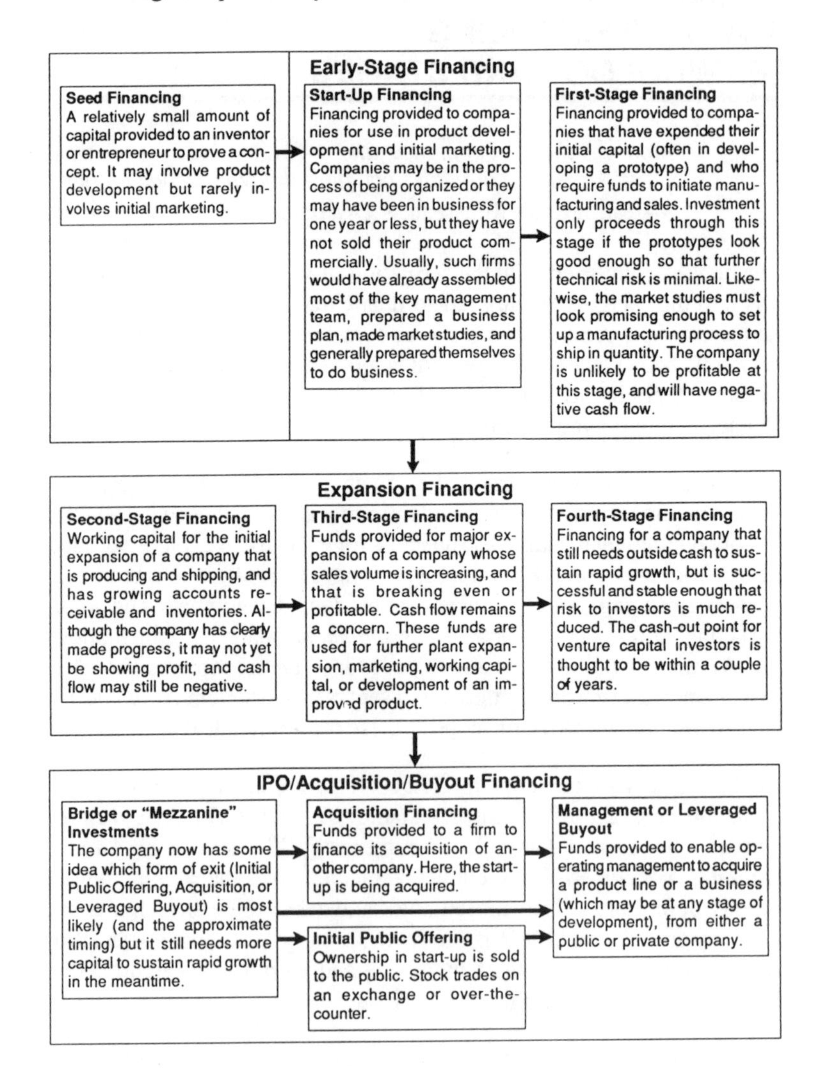

Figure 6-1 Stages of a Company's Growth

The term start-up as used by the layperson is often similarly misused by the first-time entrepreneur in his or her representations to investors. Be certain not to expose naivete in these circumstances. Likewise, the term early development is often erroneously used to refer to a pre-start-up or pre-seed-level company. If you incorrectly represent your recently incorporated but yet uncapitalized company to an investor as one in early development when it is actually in pre-seed stage, you probably just lowered your valuation by about 50% in his or her eyes! Study Figure 6-1 closely before approaching investors for funds. You must speak the language to play the start-up game. There really is not that much to learn here, but it is essential that you learn a few key terms. This chapter has been kept short so that you can study and master this vernacular.

You should be conversant with expansion financing terms and business growth milestones. Postexpansion financing terminology helps you recognize the need and understand the alternatives for cashing out when you are finished with your start-up. You will hear your investors discussing their postexpansion financing exit strategies. They will need to return to themselves, their limited partners, or other funding sources their invested capital and gains at some point in time, usually within three to seven years. During weak IPO markets (e.g., during the late 1980s into 1990), investors found themselves taking a longer ride with their private start-up investments than in the past (e.g., during the mid-1980s). During a strong IPO market, such as that experienced in 1991 and early 1992, things turn around rapidly. However, no one can predict how long the latest hot IPO market window will last.

In 1991 and early 1992, in the midst of an extended recession, a remarkable event occurred: the IPO market unexpectedly opened up. Many start-ups on the verge of bankruptcy, with nil sales and millions of dollars in accumulated losses, suddenly found themselves on the IPO bandwagon. On March 19, 1992, the *Wall Street Journal* reported that

> IPOs continue to explode with sales headed for a record first quarter. More than 120 companies have raised $7.4 billion from initial stock sales since January 1. At this rate, first quarter IPO sales are expected to be a record. If the pace continues, this year could well shatter the record $18.3 billion raised by IPOs in 1986, as well as last year's near-record $16.4 billion.

All of this was happening in the midst of one of the longest and hardest recessions in years.

Figure 6-2 illustrates the volume (in total dollars and number of issues) of IPOs during the past 12 years (adapted from *Barron's*, December 23,

1991, Wall Street's Baby Boomers: 1991 Rates as a Spectacular Year for IPOs," p. 35). The average deal size has increased over these years from about $10 million to $40 million, indicating that the more recent IPO underwritings were for stronger, larger, usually more mature companies. Many hot, high-ticket biotechnology deals helped escalate the total dollar volumes and the larger deal sizes seen in the past two years. The most common exit strategy today, however, is the acquisition of the start-up company by a larger corporation.

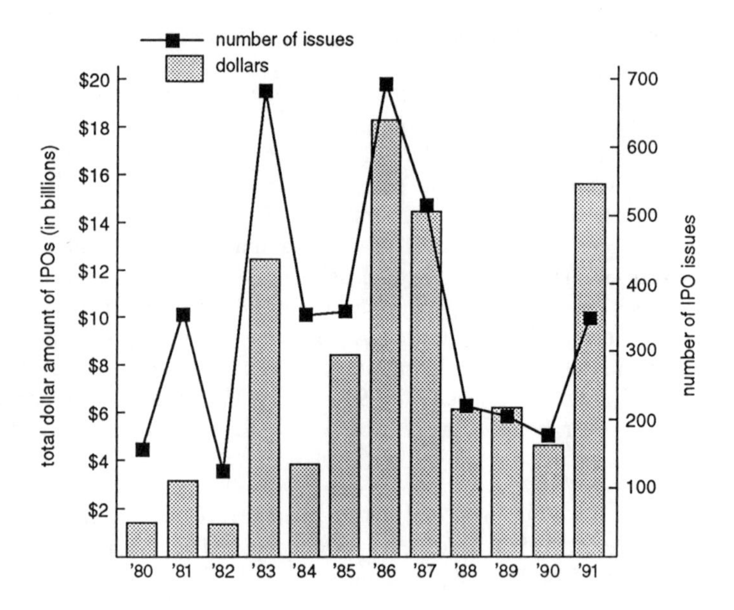

Figure 6-2 Twelve Years of IPOs

Financials for Engineers— A Crash Course

MEASURES OF FINANCIAL STABILITY

Related to raising funds is maintaining the financial stability of your company once it is launched. As you build your business, you will want to institute financial controls to make sure that your business stays healthy and does not get into trouble, and you need to know when additional funds will be required. Because it is important that you be conversant and comfortable with financial concepts when selling your business plan, this section will explain some relevant indicators of financial health, insolvency, and bankruptcy.

RATIO ANALYSIS

There are numerous key ratios that analysts use in evaluating a company's financial position. These ratios relate to:

- *balance sheet conditions*: asset evaluation, cash accounts, receivables risk, inventory risk, prepaid expenses, company investments, fixed asset analysis, intangibles such as goodwill, deferred charges, and estimated liabilities for future costs and losses
- *liquidity analysis*: cash adequacy, trend in current liabilities to total liabilities, current liabilities to stockholders equity, current liabilities to revenue, financial flexibility, funds flow evaluation, and availability and cost of financing
- *solvency analysis*: long-term fund flow, financial solvency, unrecorded assets, unrecorded liabilities, and noncurrent liabilities
- *probability of business failure:* bankruptcy prediction

Others break ratios down into liquidity, profitability, and efficiency ratios.

Since your start-up will likely have almost no sales, no profits, few assets, little inventory, no retained earning, etc., most of these traditional ratios are meaningless. The use of a few key ratios, however, might suffice for a quick "sanity check" of the possibility of going bankrupt, which is your main concern over the near term.

Glossary of Terms for Ratio Analysis

This section contains very brief and informal definitions of the basic terms needed to compute and understand our few key ratios. These terms are used in the ratio equations that follow.

- *accounts payable*: obligations to pay for goods or services that have been acquired on open account from suppliers.
- *accounts receivable*: amounts due to the company on account from customers who have bought merchandise or received services.
- *current assets*: total cash, securities, inventory, accounts receivable, etc., that can be converted into cash within one year.
- *current liabilities*: total of all monies such as accounts payable and salaries payable owed by the company that will fall due within one year.
- *net sales*: equals gross sales less sales returns and allowances, sales discounts, etc.
- *total assets:* equals current assets plus fixed assets
- working capital: a liquidity measure equal to current assets minus current liabilities. It is called working capital because it is the amount available to operate your business on a daily basis.

Key Liquidity Ratios

The *current ratio* measures your ability to meet short-term obligations.

$$\text{current ratio} = \frac{\text{current assets}}{\text{current liabilities}}$$

A good rule of thumb is that your current ratio should be greater than or equal to 2.0 (i.e., current assets should be at least twice current liabilities). If your current ratio is too low, you may not be able to pay your bills. Roberts reports that, of the high-technology companies he studied, a typical current ratio was 2.5. Companies in the electronics industry averages 3.17.

The *quick ratio* is a variation of the current ratio; it is more of an acid test of your ability to meet short-term obligations. The quick ratio eliminates inventory that is less liquid, and some people further discount accounts receivable by 25% to better reflect liquidity.

$$\text{quick ratio} = \frac{\text{cash + accounts receivable}}{\text{current liabilities}}$$

A safe quick ratio would be at least 1.0

The *turnover of cast ratio* measures the turnover of working capital to finance your sales. The ratio should be below 5 or 6. Make sure you have sufficient capital to finance your level of sales.

$$\text{turnover of cash ratio} = \frac{\text{sales}}{\text{working capital}}$$

Key Efficiency Ratio

The investment turnover ratio measures your ability to generate sales in relation to your assets, which is especially important if your business requires a large investment in fixed assets.

$$\text{investment turnover ratio} = \frac{\text{net sales}}{\text{total assets}}$$

Altman's Z-Score Measure for Predicting Bankruptcy

One very interesting computation you can try is E. Altmans *Z*-score measure of predicting bankruptcy in the short run. Auditors are required to recognize and report on possible business failure. If a business does fail and the auditor has not mentioned problems concerning the continuity of the business, a liability suit could be instituted. Don not be surprised if Altman's formula shows your start-up to have a high probability of failure—many of the numerators in the equation will be zero. At least you know where you stand; start-ups are risky business!

$$Z_{\text{score}} = \frac{\text{working capital x 1.2}}{\text{total assets}} + \frac{\text{retained earnings x 1.4}}{\text{total assets}}$$

$$+ \frac{\text{operating incomes x 3.3}}{\text{total assets}}$$

$$+ \frac{\text{marketing value of common and preferred stock x 10}}{\text{total liabilities}}$$

$$+ \frac{\text{sales x 10}}{\text{total assets}}$$

Z_{score}	probability of failure
1.8 or less	very high
1.81 or less	high
2.8 to 2.9	possible
3.0 or higher	very low

INSOLVENCY AND BANKRUPTCY

Many start-up founders will someday find their ventures to be insolvent or (technically) bankrupt. You need to know what these terms mean.

Insolvency

Insolvency may refer to either equity insolvency or bankruptcy insolvency.

Equity insolvency means that the business is unable to pay its debts as they mature. It is common for a start-up business to be equity insolvent (unable to meet its daily debts), yet have assets that exceed in value its liabilities. Such a business would be said to be *illiquid.*

Bankruptcy insolvency means that the aggregate liabilities of the business exceed its assets.

Insolvency, while not in itself proof that your business will fail, should be treated as a serious warning signal—one that can lead to bankruptcy.

Bankruptcy

Bankruptcy is a serious condition that can lead to the liquidation or reorganization of your company in order to satisfy creditor or stockholder claims.

Technical bankruptcy is the term used when a company has already committed an act of bankruptcy while insolvent, which would allow a creditor to file a court petition forcing the company into formal bankruptcy. There are six acts of bankruptcy, one as simple as giving preference to a creditor during insolvency.

In addition to periodically generating and analyzing formal cash management control documents, indicators, and ratios, you also will want to get a quick intuitive handle on your cash needs. To do this, examine how much you are spending each month, factor in any cash from sales that you might generate, and then compute how long you can last until you secure a new round of financing. Always keep your eye on that point where you could run out of money, and do your best never to let it happen.

Trials and Tribulations in the Financing of One Start-up Company

This is both a happy and a sad story. It tells of one of the more interesting small start-ups in the high-technology field. PPT (originally incorporated as Pattern Processing Corporation), based on exciting technology in an exciting field, got a great start and was supported financially for a long time despite six years of continuous losses. Two of the three founders left the company before it became profitable. New management took over to turn the company around, and the original exotic technology was set aside for a more practical variety. Included in this case study are many numbers, valuation, and percentages to give a realistic example of equity ownership, dilution, and valuations that you might expect in your start-up.

Background

In the early 1980s Larry Werth, 34, decided to quit his job at Medtronic Inc. (a high-tech medical products company in Minneapolis), in order to launch his start-up. A few years earlier Werth had done some graduate work on an interesting technical idea for robotic machine vision—giving eyes to computers to inspect industrial parts. This was exciting technology that was being eyed by venture capitalists across the country. Control Data Corporation provided a popular low-cost incubator facility for many Minneapolis start-ups in its modern downtown building. Werth rented 500 square feet in the Control Data incubator with the aid of a business assistance grant from the Small Business Association, who subsidized two-thirds of the first $10,000 of costs for business services provided to PPT by an affiliate of the Control Data Business Advisors Program.

High Technology

In the mid-1970s Werth was a research assistant in the engineering department at the University of Minnesota. There he invented a statistical pattern classification machine vision algorithm that was motivated by neurophysiology studies. His business idea was to cast this algorithm into hardware, creating unique, high-speed device to offer to the emerging machine vision market. An interesting twist was that he would not disclose the function of the algorithm to his potential customer or investors, promoting instead only its benefits when applied to solving certain problems. Werth gave little thought to the potential market for his invention. However, members of the investment community were soon stumbling over each other looking for deals in the new and exotic machine vision and robotics industry.

Seed Financing

Werth wrote a partial business plan while employed at Medtronic and told several investors the story of his invention and his passion to build a business around it. The business was incorporated December 9, 1981. For the next three months, while employed at Medtronic, Werth and company searched for capital and researched the availability of patent protection.

A few months later, two venture capital firms seeded the business with a total of $60,000 (enough to carry Werth's team for about six months) in exchange for 15.6% of the company. This valued PPT at $385,000 on a post money basis. As soon as this seed capital was assured, Werth and his two cofounders quit their jobs a Medtronic.

Werth's start-up team consisted of himself, Mike Haider (a financial specialist who also led marketing), and Larry Paulson, a sharp hardware design engineer. Werth's lean team, representing the bare necessities, was quite attractive, containing leadership, marketing, and quality engineering. Commencing April 1, 1982, Werth, Paulson, and Haider began receiving $35,000, annual salaries.

Start-Up and IPO at the Same Time

Because a new securities law was generated that permitted the creation of quick, small intrastate public stock offerings, in early 1983, less than one year after start-up, Werth and his team were members of a publicly-traded corporation. Pattern Processing Technologies' IPO raised $300,000 in exchange for 44% of the stock, for a post-offering valuation of about $650,000.

Early Development Financing

PPT soon thereafter raised its total equity financing to $1.7 million.

Just before the IPO, Werth owned 129,600 shares (33.8%) at a cost of $800, Haider and Paulson each owned 97,200 shares (25.3%) at a cost of $600, and the two small venture firms owned 30,000 shares each (7.8%) with warrants to purchase an additional 40,000 shares each, for a total cost of $60,000. The founders' shares cost them about six-tenths of a cent per share (adjusted for splits and stock dividends.)

Expansion Financing

PPT had raised over $10.5 million in private and public equity financing through 1991. Its stock price was published in the paper daily, and it did quite well from 1983 to 1987, rising from $1,00 to about $10.00. After six years of continued nonprofitability, from 1983-1989, however, the stock was reverse split in 1989 into one share for every 20. Share sold for well over $2.00 in 1992, which is equivalent to about $0.10 adjusted for splits. PPT was not profitable until fiscal year 1990, and the founders'' shares were substantially diluted in the meantime. Stockholders made money in late 1989 to 1991, as the stock rose from $0.50 to over $3.00 when the company finally made the transition to profitability on a quarterly annualized sales rate of $2.6 million. With just under 2 million shares outstanding on a fully diluted basis, this small company of 25 employees has a tidy market capitalization of about $5 million on sales of about $1.8 million. It is worth noting that the company net operating loss (NOL) carryforwards for tax purposes of about $8 million. The value of this company for its NOL tax write-off potential alone would be worth about $4 million to an acquiring business in the 50% tax bracket.

Postscript

PPT no longer utilizes the original technology upon which the company was based, and is one of the last surviving companies in the machine vision business. In 1989, Werth moved on to pursue his pattern-recognition idea at Electro-Sensors, Inc., with the backing of a related investor group. Werth notes,

> There is still value in the idea of casting algorithms into hardware to realize high speed. My technology was important then and it contributed to helping PPT survive while most others in that business failed.

Werth is right, and you want to make sure that you have technology-fuel for your business. Werth came out of the experience whole—not rich, not impoverished. Werth suggests,

> Don't underestimate the slow pace at which industry makes purchases [of high-technology products]. Promotion alone won't change the fact.

This time he is doing his research and development first and focusing on a specific application (market need): computerized reading of paper forms filled out by hand.

In conclusion, if you too have a good idea, a little luck in finding funds, a willingness to leave your present job, and the passion and conviction to start your own business, the preceding story could be about you. Becoming more successful in your business than the PPT founding team, however, requires a more focused market- and customer-driven technology-fueled strategy.

You cannot go public as a start-up today unless you find yourself in an exceptional situation. You might, however, be able to go public in an active IPO market after your investors have invested several million dollars and if you have a new product with prospects for sales, even if you are toying with insolvency and have nil sales to date.

Finally, be cautious of inventing on company time and with company resources. If Werth's invention had been of any interest to Medtronic, he could have had a nasty intellectual property rights challenge on his hands.

Part Three

DUE DILIGENCE

Part Three of this book really gets into the meat of things. Before you launch your business you need to thoroughly understand the basic ingredients of a successful start-up. Because these topics are to important, Chapter 7 is devoted solely to outlining them. You will need a very clear map of where you must travel if you ever expect to get a satisfactory destination.

The process of investigating your start-up's success factors is called *due diligence*. Before investors put cash into a start-up, they exercise due diligence to try to discover everything that could impact their business investment. You must do likewise. Investigate and master all the success-contribution factors that will make or break your business; management team, board of directors, markets, customers, products, business plans, and fund-raising. These form the base upon which you will build your business.

David H. Bowen, publisher of *Software Success*, emphasizes,

> Investors look at between ten and one thousand possible investments for every one they make! Entrepreneurs make very real investments of their sweat and years of their life. Don't start a company just to do it. Ask yourself if you would invest as your alter ego. Don't start a company you wouldn't invest in.

Chapter 7
ELEMENTS OF A SUCCESSFUL START-UP

"Conventional wisdom: Get a million-dollar idea, find some venture capital and go. Reality: Venture capital companies are not interested in ideas. Get some seed money, make sure your prototype and your company are 'debugged,' and then go."
—Gordon B. Baty

SUCCESS INGREDIENTS

This chapter gives a brief introduction to the following five chapters which comprise the meat of the book.

What is the first thing a company has to have to be in business? It may seem like a cliche, but not everyone knows the answer. You need a customer, which relates to the broader concept of a market to sell to. If no one purchases your product or service, you will soon go broke. This leads to the suggestion that your start-up should also have a product or service, or at least the technological base and firm plans to develop one. Obviously, you will need management talent, too, to run your business efficiently, and even the best management will fail without adequate financing. Finally, every company needs a business plan; otherwise you will not know what you and your competitors are doing, where you want to go, how you will get there, and when you have reached success.

These critical ingredients, being tightly connected, cannot be thoroughly examined out of context. Figure 7-1 illustrates the relative importance of each element for a typical start-up situation. As for other ingredients for success, such as luck and persistence, you are on your own.

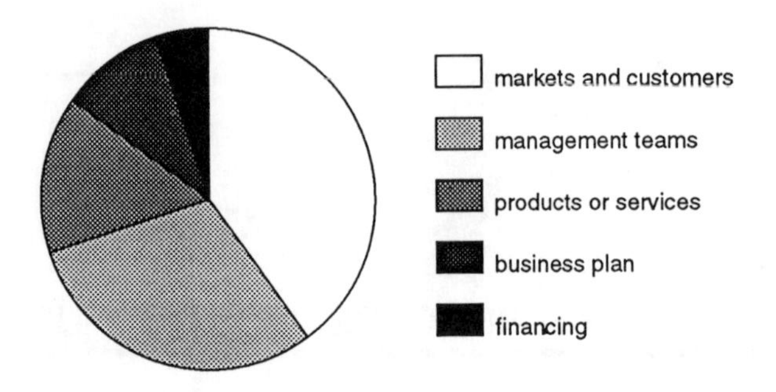

Figure 7-1 Five Controllable Ingredients for Start-Up Success

Because it is so important for the entrepreneurial engineer to understand what makes a business successful, a separate chapter is devoted to each of the five principal start-up elements.

LEADERSHIP AND BUSINESS BASICS

Success derives from the disciplined administration of a written or unwritten plan for coordinating and leading the energies and resources of a variety of players toward a common vision and mission. The founding entrepreneur must provide this broad vision and mission, and demonstrate the leadership needed to grow the company. While each CEO will have a particular area of interest or expertise, such as selling or developing a product, he or she must be able to operate across the broad spectrum.

For the start-up entrepreneur, Steven Brandt's *Entrepreneuring: The Ten Commandments for Building a Growth Company* lists 10 important operational leadership style-related activities you must execute consistently and well. His 10 "commandments" are shown in the left-hand column of Table 7-1.

The center column lists associated classical management functions and the right-hand column lists related points that are emphasized in this book. Understanding the relationships between these business basics will help you to develop your own effective leadership style.

Table 7-1 Entrepreneurial Success Through Classical Management Functions

BRANDT'S COMMANDMENTS	CLASSICAL MANAGEMENT FUNCTION	EMPAHSIS FOR THE ENTREPRENEUR
1. Limit the number of primary participants to people who can conciously agree upon and directly contribute to that which the ententerprise is to accomplish, for whom, and by when	staffing	Launch your start-up with a complete, experienced, and compatible management team.
2. Define the business of the enterprise in terms of what is to be bought, precisely by whom (i.e., the customers), and why.	planning	Use a market- and customer-driven strategy to define your product.
3. Concentrate all available resources on accomplishing two or three specific operational objectives within a given period of time.	organizational directing controlling	A superb business plan calls for a superb, focused execution.
4. Prepare and work from a written plan that delineates who in the total organization is to do what, by when.	planning	Write a solid business plan that the team believes in.
5. Employ key people with proven records of success at doing what needs to be done in a manner consistent with the desired value system of the enterprise.	staffing	Create a complete, experienced, and compatible management team.
6. Reward individual performance that exceeds agreed-upon standards.	staffing	Motivate with a fair, remuneration plan, including equity participation.
7. Expand methodically from a profitable base toward a balanced business.	controlling	Pursue rapid profitability leading to high growth.
8. Project, monitor, and conserve cash and credit capability.	controlling	Never run out of money!
9. Maintain a detached point of view.	planning	Develop a market-driven strategy. Do not concentrate on your technology to the exclusion of other success factors
10. Anticipate incessant change by periodically testing adopted business plans for their consistency with the realities of the world marketplace.	planning controlling	Develop and maintain an operational business plan after funding is obtained.

* Adapted from Steven C. Brandt, Entrepreneuring: The Ten Commandments for Building a Growth Company, New York: New American Library, 1982, with permission of the author.

Chapter 8
CREATE YOUR MANAGEMENT TEAM AND BOARD OF DIRECTORS

"There is something rarer than ability. It is the ability to recognize ability."
—Elbert Hubbard

MANAGEMENT

An old cliche of real estate is that the three most important factors in selling a house are location, location, and location. Arthur Rock, a venture capitalist, was quoted as saying that, in starting a new company, the three most important factors are "people, people, and people." As is discussed throughout this book, there are several dimensions for success in selling your business, including people (management), markets and customers, products and technology, planning, and financing. This chapter focuses on the people (namely management teams), as well as your board of directors.

	inexperienced (0)	experienced (1)	very experienced (2)
complete team (2)	2	3	4
partial team (1)	1	2	3
no team (0)	0	1	2

Figure 8.1 Management Completeness-Experience Grid

You will start your business somewhere in the management completeness-experience grid depicted in Figure 8.1. Will you start your business with a full or a partial management team? Will that team be experienced or inexperienced?

The value of your management team in the eyes of investors will be directly related to the experience and completeness of your team. One simple metric is obtained by assigning zero, one, or two points each for completeness and experience. The sum then indicates the overall perceived capability of your start-up's management to investors and customers. Which companies in the grid would you want to invest in or buy a product from—the 1s, 2s, 3s, or 4s?

It has been established that in order to achieve high growth and to be successful in the long run, you will want to grow your business with a top-notch management team. There are many examples of successful start-ups founded by inexperienced management (Apple Computer, Inc., Sun Microsystems, etc.) who either turned out to be top-notch managers or had to be replaced. The key is to identify when and if one needs to bring in more experience. In particular, you will have to decide for yourself how important and practical it is for you to be or remain the CEO.

Though a good market is critical also, without an exceptional management team even the most ideal market cannot be fully exploited. This is evidenced in the fact that many business plans with no product and no technical team have attracted financing when written by a proven management team, and were targeted toward a growing market, which obviously had customers.

Management Status → most desirable

Management Status				
Level 4: All members on board and experienced.	5	6	7	8
Level 3: All members identified; some on board only after funding.	4	5	6	7
Level 2: Two founders; others not identified.	3	4	5	6
Level 1: Single entrepreneur.	2	3	4	5
	Level 1: Idea only; market assumed.	Level 2: Prototype operable but not developed for production; market assumed.	Level 3: Product fully developed; few or no users; market assumed.	Level 4: Product fully developed; satisfied users; market established.

Product Status

Figure 8.2 Team Size and Product Status in Business Plan Reception

The size and commitment of your management team will have a lot to do with the reception your business plan receives in the investment community. Stanley R. Rich and David E. Gumpert's *Business Plans That Win* contains an evaluation system that captures the fact that the quality and completeness of the management team and product are important to investors. Figure 8.2, adapted from Rich and Gumpert, makes it obvious that investors prefer strong, complete management teams.

Venture capital investors would like to build their start-up portfolios based on 6s and 7s. Less probable are the companies that evaluate to 4s and 5s.

MIPS Computer, the subject of Malone's book, *Going Public—MIPS Computer and the Entrepreneurial Dream*, was known as the $100 million company with the $1 billion over-the-hill-gang management team. Largely because of its solid management team, MIPS was one of the first high-technology companies able to go public after the 1987 stock market crash.

Most start-ups are founded by people much like yourself. These founders have ambition, persistence, good ideas, and an ability to work with and motivate people. They may also have financial savvy. Often founders are persistent to a point of obsession. Unfortunately, though, some founders may retain all control, give poor direction, make business decisions for nonbusiness reasons, bring in ineffective relatives or friends, and be slow to correct operational problems. Worst of all, they may not be capable of learning from the past, thus tending to repeat mistakes. Will your start-up team have perfect, proven management? Probably not initially. It is an imperfect world. How do you go about selecting quality team members?

A Kind of Marriage

The management team of a start-up will likely share more time and experiences, solve more problems, and ride rougher waters together than any other association of individuals. It really is a kind of marriage. Few secrets will or can be kept, and common interests and visions drive the survival of the relationship. Deep feelings from respect to hatred are likely to develop. Like marriages, many relationships will dissolve. It is vitally important that you choose the right team members at the beginning. Do not launch your start-up with any less due diligence than you would for a marriage!

If possible, consult or work part time with candidates before making any decisions. Get to know the people. Make sure they pass the chemistry test. If you like a good laugh now and then, but a candidate never cracks a smile, you two might not make a very happy team.

Team Members

Your core start-up team will probably consist of three members:

- team leader (presumably you, holding the CEO and president titles)
- vice president of marketing and sales
- vice president of engineering (perhaps also holding the chief technical officer [CTO] title)

Together, this team will drive the business and determine what product to develop and how to build and sell it.

Reasonably soon you will need to add, at least part time, a chief financial officer (CFO) for financial controls and a vice president of manufacturing if your product has significant manufacturing content (unlike software).

A glance at this list will emphasize that you, as CEO, will have far too many duties to manage the entire business and manage the engineering development activity also. Many start-up entrepreneurs cannot let go of the engineering roles with which they have been associated and in which they find their primary strengths. You will have to make the decision to:

- be the CEO and leave the engineering to someone else
- let one of your team members take the CEO position
- possibly compromise the initial growth of your business by acting as both the CEO and the engineer

Many entrepreneurs might argue for the last option, pointing out many success stories. However, in almost all of these cases, initial growth was limited, impacting the rapid creation of wealth. Also, the businesses were eventually successful, usually because the entrepreneur was able to grow to assume the CEO role full time, successfully replacing his or her part-time engineering position with a full-time engineering manager, or the entrepreneur eventually relinquished the role of CEO, taking on the engineering management position full time.

Know Yourself

Before jumping ahead to build your winning team, you first need to get to know yourself. Dr. Philip B. Nelson of the Institute for Exceptional Performance is frequently called upon by top executive recruiting firms to measure the characteristics and competencies of candidates, compare that data with the desired position characteristics and competencies, and recommend individuals who would be compatible and synergistic in top-performing management teams. He utilizes a proprietary Position Suitability Profile SystemsTM worksheet to plot the following characteristics and competencies:

- problem solving: thoroughness, practicality, analytical ability, creativity, broad perspective
- motivation: drive, determination, persistence, initiative, goal orientation
- work habits: self-discipline, responsibility, decisiveness, integrity, dependability
- organization/planning: planning, organization, setting priorities, punctuality, flexibility
- interpersonal characteristics: self-confidence, amiability, persuasiveness, stability, perceptiveness
- leadership characteristics: delegation, firmness, participation, recognition, example

Not every reader will have access to Dr. Nelson, but you can achieve similar self-analysis results with a little work.

- Get candid feedback from your peers on how you appear to them.
- Decide on the types of team members you will need to complement your skills, challenge you to do your best, and supplement your weaknesses.
- If you bring together team members with whom you are most comfortable, you may not end up with a well-balanced team. For example, if you are not a thorough person, but you feel most comfortable with similar people, who on your team will be thorough enough when you need it?
- Use a matrix to plot some of Nelson's characteristics, styles, and competencies of potential team members. Look for dangerous similarities or extreme incompatibilities with your personality.

A Winning Team

Now that you have a clearer view of your own strengths and weaknesses, you can visualize what characteristics and competencies in others will round out and complement your team. High-performing management teams must be compatible and synergistic. Each member must:

- challenge the others
- provide mutual inspiration
- get along and work well together
- be able to perform in contained chaos
- maintain control despite the extreme pressure

One profile of successful winning start-up presidents (recounted by Charles A. Skorina, an executive search consultant with Charles A. Skorina & Co.), provides a three-point checklist, useful not only for yourself but for your potential team members as well. Winners:

1. thrive on risk
2. are incurable optimists
3. have dogged persistence

Make sure both you and your team will thrive on the chaos often associated with a start-up. In all likelihood, at some point in time, key employees will leave, prototypes will fail, money will run out, or key customers will vanish. You must be able to stay the course, pursue the goal, and enjoy the game.

Successful Matches

Do more than just hope for smooth teamwork. If prolonged disputes or shouting matches occur in the frenzy of your start-up, it will be destructive of morale and performance. Look for good matches. Avoid the following incompatible differences in work habits and ethics, too many of which may signal danger in your proposed relationship.

- small company orientation vs. big company orientation
- open and generous vs. protective
- sense of humor vs. humorless
- high energy vs. low energy
- team player vs. individual player
- honest and direct vs. cagey and indirect
- sees glass half-full vs. sees glass half-empty
- treats others with respect vs. treats others as objects

If the values, goals, and objectives of you and a proposed partner do not mesh, then it would be wise not to join together in this start-up. Assuming you have identified no serious conflicts, teaming with otherwise qualified old friends may work well if you can establish a decision-making process that works smoothly.

Signing on Management Team Members

You do not simply interview people to select them for inclusion in your team. You need to develop trust and respect in a relationship, and you must share a common vision of what the business can and should be. It takes time to find and meet the right people, and this step in growing your business is critical.

If you select even one poor team member, your venture is highly likely to fail. A start-up generally does not have the luxury to make such a costly mistake. Besides, beyond creating wealth, you will want more than all else to make this an enjoyable adventure.

Make sure that you understand your candidates' backgrounds: where they have succeeded and failed, what they have learned, how they intend to help run this business differently, etc. Ask for references of past employers and coworkers, and speak with them regarding the personalities and work habits of the candidates.

If you think an unproven management candidate has potential, will that person be willing to step down to replace himself or herself with more professional management if need be as the business grows? Will your potential team members agree to move aside if proven to be less than effective? You must create a team with a balance between doing and managing that will result in action. As technically oriented and creative managers, you and your team members must, if you discover an inability to manage among yourselves, be the first to suggest hiring your own replacements. Ask people such questions directly, though, and you may be surprised at the variety of answers you will hear. Ask, and then listen carefully to the answers. Do not rely on the assumption that your management candidates must be good to have gotten where they are.

THE ENTREPRENEURIAL TEAM

Bob Hansens, president of the Silicon Valley Entrepreneurs Club, artfully describes the entrepreneurial team members and their related roles. It may be useful for you to think of your start-up team in terms of his structure, which is slightly modified and transcribed in Table 8.1.

Table 8.1 The Entrepreneurial Team

Team Leaders	Achievement-Oriented Managers	Technology Team Leaders	Advisory Board (part time)
chairman of the board	chief operating officer (COO)	chief technical officer (CTO)	entrepreneurial team members
chief executive officer (CEO)	chief financial officer (CFO)	vice president of engineering	providers of professional services (law, accounting, etc.)
president	vice president of marketing	director of technology	providers of capital
	vice president of sales	chief scientist	industry experts and consultants
	vice president of manufacturing		university professors

As described in Chapter 2, one individual can hold multiple titles and positions. The CEO is the most important and, subsequently, potentially the weakest link in a start-up. As the CEO, you must be able to manage teams of people who are difficult to manage.

Your Key Employee Team

As you begin adding key technical employees to your payroll, you will be looking primarily for technical competence. However, do not overlook that team-player quality that you sought in your management team. You will want the entire enterprise to share the vision and live the mission of the business.

Two qualities you should require are energy and enthusiasm. Employees with these attributes know no limits, think that they can do anything, and often come pretty close to doing so in practice. Especially amazing is the extreme productivity of some of the younger and less experienced employees who often exhibit this pure energy and enthusiasm. These individuals, usually hired at relatively low entry-level salaries, often rapidly become key employees in a business, and they deserve to be treated and financially rewarded as such. Do not hire solely on the basis of age, however. Age discrimination is illegal, and clearly not all young people are energetic and enthusiastic. Many older individuals still have that energy and enthusiasm you are looking for, in addition to their valuable experience. Also, while many young free spirits may produce terrific results for you in the short term, their long-term interests may reside elsewhere. The turnover of new college graduates (those holding their first jobs) can be quite high.

BOARD OF DIRECTORS

Power of the Board

The board of directors determines how much you will be paid and has the power to replace you as the CEO, so it is natural that you would want to control who gets on your board. When your company is first incorporated and before it is capitalized, you are the only significant shareholder and can appoint whoever you want. In California, you can be the entire board of directors by yourself if you so desire.

A board of directors is elected by the shareholders of the company and is empowered to carry out certain tasks as spelled out in the corporation's charter. Among such powers are: appointing senior management, naming members of executive and finance committees (if any), issuing additional shares, and declaring any dividends. Boards normally include the top corporate executives (the inside directors), as well as outside directors chosen from both the business community and the community at large to advise on matters of broad policy. You should discuss with your advisors whether only you as the CEO, or other members of your management team as well, should be on the board.

What a Board Should Do

The best-utilized boards will provide objectivity and sound judgment. Directors will question your assumptions, contribute to the resolution of specific problems, and bring new ideas to the table. Expect the board to review financial performance, marketing plans, key hiring decisions, and any other major development affecting the business.

How Often Does a Board Meet?

Venture-backed start-up boards usually try to meet every four, six, or eight weeks, or quarterly, but this varies significantly.

How Long Does a Board Meet?

Board meetings should last an hour and a half or, at most, two hours, but many run the better part of a day in start-up situations when many issues, including operational matters, are to be discussed.

Selection of Directors

Because they generally want to retain some control over their investments, you will probably end up having investors on your board. It is generally wise to get concerned board directors, and your investors will certainly fall into that category. Whether or not they will have too much vested interest in looking out for their finances is pertinent. A typical Silicon Valley start-up board comprises only the funding venture capitalists and the CEO.

Investors regularly demand seats on the board as a condition for investing, even if they do not have a majority of shares to elect themselves on board. For example, it is not at all unusual for investors' preferred series shares to have expanded voting rights over the common shareholders to facilitate their own board membership and increased control of the board.

If you have investors on your board, try to get individuals who are experienced in your industry and can genuinely help. These people will have run companies themselves. You will want someone on your board who understands financial affairs and the problems of running a business profitably. You should find individuals with sound judgment and who hold positions of leadership in your field. All directors should be relatively free from conflicts of interest and be able and willing to devote the time required.

It is also recommended that you strive for diversity on your board. Areas of expertise that are most valued are legal, finance, management, marketing, and human resources. An ideal board for a high-technology firm might consist of a venture capital firm partner, a university president, a *Fortune* 500 company financial strategist, an operations manager from your industry, and a retired CEO of a related company.

The board of directors is important in the long run, and having a supportive and knowledgeable board is crucial to the success of a company and the well-being of its founders.

Edward Roberts of MIT, in examining high-technology companies, found that boards typically comprise six to seven members (both outside and inside) with a variety of backgrounds. Representative professions prevalent on boards were, in order of frequency:

1. finance (venture capital, banking, private investing)
2. in-house
3. business (company-related, general)

4. consulting
5. academia
6. law

In addition, you should select directors who treat others the way you would like to be treated. For example, in one case where the CEO had to be replaced, the investor board member insisted that the company provide this individual with an office and secretarial support for the better part of a year, provide six months' severance pay, and allow the CEO to continue vesting in his stock during the severance period. One objective was to make the involuntary parting as amicable as it could be. They wanted the outgoing CEO to sincerely speak well of the company and its backers. That is the action of a first-class investor and director who is thinking of the longer term.

A board of directors can be a very powerful source of ideas, guidance, and leadership. The wrong board can be a nightmare. Most boards prefer to let the CEO run the business, and only step in on a more frequent basis when they see problems. As CEO, you should be on the board, and you might very well occupy the position as chairman of the board. Practically speaking, however, the chairman position carries little additional power, especially if you are already the CEO.

How Many Members Should Be on Your Board?

A start-up company can do well with a small board. A good recommendation would be to have between four and six directors. Janet G. Effland of the venture capital firm Patricof & Co. Ventures, Inc., prefers to see five or seven since with an odd number there is always resolution of a controversial subject. One workable combination for your initial board would be:

- you
- an outside financial advisor (perhaps your part-time CFO)
- your first-round investor
- a highly respected business advisor (a potential second-round investor would be ideal, but be aware that most active investors do not have time for noninvestee boards)

Compensating the Board

If a board member is a significant shareholder in the company, no compensation is required. He or she will have the motivation to be active, be attentive to the business, carefully review documentation, attend meetings, and perform committee work. If a director is not a stockholder, you will want to offer cash compensation or stock options (or possibly both, though you may only be able to afford the options in your start-up phase). Reimburse your directors in phases to make sure that those who stick with you get their rewards. While people will often serve on boards of large, prestigious companies for little or no compensation, more than likely you will need to compensate your nonstockholder directors. The American Electronics Association's 1991 *Executive Compensation in the Electronics Industry* survey publication reports that of its 423 reporting private companies, 379 (90%) provided no compensation for inside directors. Fifty-two percent provided no compensation for outside directors. Even when compensation was provided for outside directors, the average annual retainer was only $5,430. Twenty-seven percent provided stock benefits. AEA private companies include some pretty large businesses, along with some smaller start-ups, so compensation for the board of your start-up will probably fall below the AEA average. In contrast, AEA public companies provided stock benefits to 61.3% of their board members.

Management of the Board

Primarily, your management of the board will consist of making sure that it is composed of competent people who will put the time and energy into helping your enterprise. You do not let directors run the business on a day-to-day basis unless a problem arises.

You should also set the tone for board meetings and try to organize the agenda to focus on crucial issues. It helps to give assignments to directors for them to complete prior to the next board meeting. Do not turn the director's meeting strictly into a show-and-tell for your investors.

Prepare for Meetings

By properly preparing for board meetings, you will help to manage the board more efficiently. Providing updated financial statements and updates on major developments and carefully selecting the issues to bring to the board's attention will assist you in thinking through your business.

Before a board meeting, consider each issue and the options involved. Know what actions you want from your board, and let them know their part up front.

It helps to get premeeting information to your directors a minimum of two days before the meeting so they can be prepared. Keep in mind also that your company's attorney (often a difficult person to schedule) often acts as secretary to the board.

Legal Liability for Directors

Since boards are legally responsible for the actions of the corporation, many people will not formally serve due to legitimate concerns for liability. While a 1988 California law enables companies to reduce exposure to board members with an appropriate amendment to the articles of incorporation, members still do remain somewhat exposed. Purchasing director's liability insurance, while it sounds like a good idea, will most likely be beyond your initial financial capability. (You will need it, however, one year before your IPO.) The expense of liability insurance is reflected in AEA statistics that show that only 8.3% of private AEA members provide liability insurance. Consult your legal counsel to determine how you should best address this issue in your start-up.

ADVISORY BOARD

Many companies form executive advisory boards or scientific advisory boards to bring additional consulting expertise into the company at low cost. Because these boards have no legal purpose they invoke very limited legal liability exposure, and many individuals would be honored to be included in such a manner. These experts can be quite useful as technical consultants to your company, and if properly motivated (make them stockholders), they can do wonders for your image in the industry and trade press.

Mentors

If you have managed to form a mentor relationship during your career, be sure to include your mentor in your start-up business activities, perhaps as chairperson of an advisory board. This is the time to reward such a valuable individual with some stock options and to keep him or her interested and involved in what you are doing.

Chapter 9
EVALUATE
MARKETS AND TARGET CUSTOMERS

"It is better to select an audience and fit products to it than find a product and fit an audience to it."
—Freeman F. Gosden, Jr.

TRADITIONAL BUSINESS MODEL

The traditional business model strategy for starting a firm suggests that you:

- Stake out a niche market in which you will have dominance from the start.
- Serve your market through increasingly better customer service and support.
- Develop a refined grasp of your market's distribution channels.
- Create product identity, company identity, and customer loyalty.
- Establish insurmountable barriers to entry by competitors.

Your first task, then, is to establish a market niche in which your technology and product can profitably play. To do that, you need to know your customers and markets.

CUSTOMERS AND MARKETS

Customers

When launching your start-up, you must have at least a vague idea of who your first customers could be. Your persistence and enthusiasm do not guarantee that these customers really exist. Some start-up entrepreneurs have no idea to whom they will sell their first product or service, what the customer will pay, what problems will be solved for the customer, or what the customer's alternatives are. Even worse, more than one entrepreneur has believed that everyone will want to buy his product. Not since *Life* magazine has anyone sold anything to everyone. Today, markets are more specialized; customers demand products and services that will deliver to them exactly the benefits that they require. If your technology is going to result in a product that is going to be successful, you must identify a realistic, specific target audience—a niche. You must have a crystal clear vision of your customer. Sandra L. Kurtzig of ASK Computer said it well:

> One sobering aspect of choosing ASK's market was the ealization that we couldn't be all things to all people.

Know Your Customers

Before launching your company you should get to know some of your customers personally. Talk to several potential customers about your planned product or service, and emphasize that you want to know how you can help solve his or her problems. One hint: do not sell too hard. Let the customer do most of the talking (if he or she will), and listen. Ask for elaboration on any point on which he or she seems hesitant or searches for words to describe. Repeat what you believe the potential customer said in your own words: "If I understand you correctly, you feel.." Then listen again for a confirmation or clarification.

Some general questions to ask are:

- What problems do you need solutions for?
- Why does our product appeal or not appeal to you?
- Is there anything you see that is unique about our product?
- What should this product cost, and how could you justify the investment to purchase it?
- When would you need delivery? Would you be interested in acting as a beta test site (for preproduction evaluation)?
- How will your needs change in the future?
- What are your concerns or worries about our product?
- Will you be a repeat buyer if you are satisfied, or would you only need one of these products?
- What are your alternatives if you do not buy our product?

Markets

Beyond examining individual customers, you will want to look at the overall market characteristics for the product or service being offered.

- Is it a high-growth market?
- What are the market opportunities and risks?
- Are there many competitors in this market?
- Does your company have a market niche to itself?
- Is your business in a crowded market with numerous well-established competitors (for example, hard disk drives for personal computers)?
- Will your market endure over time?
- What are the risk factors for failing?

These critical customer- and market-related issues are addressed in more detail in Chapter 11. Professional investors, such as venture capitalists, will especially scrutinize market issues as part of their due diligence process prior to committing any funds to your start-up. They are probably in a better position to perform such evaluations than you are since they have access to expensive market research studies and industry association reports (not to mention their extensive network and investment experience). If you can confer with your start-up's potential investors early on, you may uncover some very pertinent market information.

Competitive Market Analysis

In addition to identifying a healthy, growing market, you will want to carefully evaluate your competition within to make sure you have a perceived distinctive competence. Figure 9.1 illustrates most of the important questions you should be asking as you evaluate different markets or analyze how a particular product idea might work.

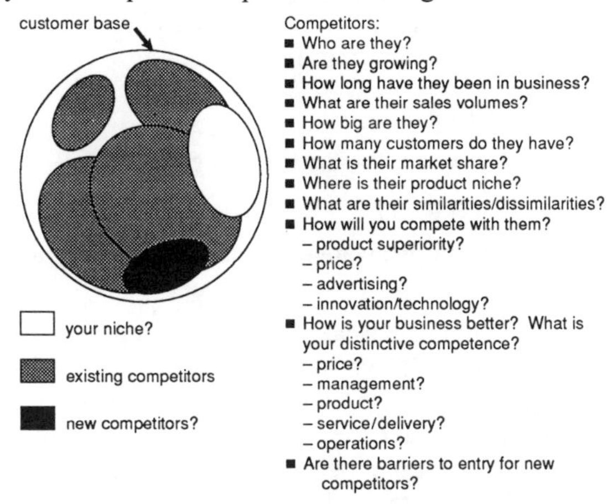

Figure 9.1 Competitive Forces in Your Marketplace

Your business plan, discussed at length in Chapter 11, will contain the results of your analysis. You will need to be able to give an account of other industry participants and prove that you either have a distinct competitive edge or a viable niche to yourself. You should compile separate profiles of each competitor (keep a separate notebook for each, noting niches served, market share, etc.) and a tabular comparison of the strengths and weaknesses of products and companies.

In your analysis, do not:

- assume there is no competition
- miss any major players
- underestimate the strength of competitors

On the other hand, you should:

- be aware of competitors' product plans and market strategies
- develop your own marketing strategy for counteracting existing and new competitors
- make perfectly clear to yourself and to your investors (in your business plan) what exactly will make your product better (i.e., how you will compete)

Marketing versus Sales

Many people confuse marketing with sales. *Sales* is dealing directly with customers and is a developed art form. *Marketing* is enticing customers to consider buying a product and is an acquired discipline. More broadly, marketing is characterized by the "four P's":

- product (what to sell)
- place (where and how to sell—distribution channels)
- price (how much to sell for)
- promotion (how to raise awareness, gain acceptance, and make people want to buy)

Marketing decisions determine what products a company is going to develop and sell, how they will be positioned, to whom they will be sold, how they will be priced and distributed, and how their existence and features will be communicated to the market. Having the right products at the right price and the right programs to effectively and profitably sell those products is fundamental to any business.

Marketing Strategy

How are you going to enter the market, obtain a niche, maintain a market share, and achieve your stated financial projections? Figure 9.2 illustrates the infrastructure you will need to build to support your marketing goals. Taken as a whole, this will be your marketing strategy, and it will find its way into your business plan.

Figure 9.2 Marketing Strategy

Based on your analysis of Figure 9.2, some questions you need to ask are:

- What is the sales appeal (special qualities or uniqueness) of your product?
- How will you attract, maintain, and expand your market?
- What are your priorities? Again, do not think that everyone is your customer. You cannot be everything to everyone.
- How will you reach decision makers?
- Will you sell via salespeople, in-house staff or manufacturing representatives, direct mail, telemarketing, or trade shows?
- To whom will you sell—value-added resellers (VARs), systems integrators, or original equipment manufacturers (OEMs)?
- How, where, and when will you advertise—in trade journals, in magazines, or on television?
- How will you generate leads? How many leads are needed per sale? What is the cost per lead?
- How long is your selling cycle?

- How big will your orders be?
- How will you pay commissions?
- In what areas of the country will you concentrate?
- What are your prices, volume discounts, and dealer discounts? Will you be involved in price wars?
- How will you package and present your product?
- How will you collect your receivables?
- What level of service, warranties, and guarantees need to be offered?
- As you can see, the marketing side of your business will consume a lot of your time. You can hire a marketing manager to take some responsibility in these areas, but as the CEO you have the ultimate obligation to make sure your business can sell, as well as make, an exceptional product.

Marketing Musts

Here are a few marketing musts.

- Do not underestimate the time required to establish a network for sales and distribution.
- Do not attempt to fill too many unrelated market gaps.
- Do not try to justify customer price by the cost to produce.
- Do justify price by the value to your customer.
- Do promote the marketable differences of your product.
- Do develop attractive packaging and establish brand recognition and loyalty among your customers.

Chapter 10
DEFINE YOUR PRODUCT OR SERVICE

"A rat trap? A rat trap? I thought you said you wanted a cat trap!"
—Terry Furtado,
Mechatronic Technologies, Inc.

OVERVIEW

Almost everyone, especially an engineer, figures they need to invent that perfect new product idea to start a business. This is not so! The sections on market pull and technology push in Chapter 4 emphasize that you should strive for superb execution in developing and marketing a perhaps less-than-exceptional product for which there exists a market, rather than a less-than-exceptional execution of a superb product for which there may be no customers without missionary sales efforts.

The purpose of this chapter is to understand a little bit more about:

- choosing the right product for your start-up
- marketing and competitive-analysis considerations in light of your chosen product
- exceptional product attributes
- producing your product

CHOOSING THE RIGHT PRODUCT

Products or Services?

You will note that this section discusses only products. This is because in a sense a service is also a product. Ideally, if your business provides a service, it can be replicated and marketed much like a product, and, also like a product, it can represent a high-growth, high-profit-margin opportunity.

The term *growth* is the key word here and was discussed in detail in Chapter 5. A service business is a perfectly acceptable way to create wealth so long as it is a growth business. Many franchises fall into the category of service businesses. It follows, then, that the way you make money is not to buy franchises, but to sell them.

Assuming you have selected a product upon which to start your business, do not forget the concept of service altogether. However, service for your product is not to be overlooked. Good service is an extremely powerful differentiator, as Davidow emphasizes in *Marketing High Technology*.

> The key is to convert great devices into great products. When a device is properly augmented [with service] so that it can be easily sold and used by a customer it becomes a product.

It is worth noting that very successful companies such as IBM have knowingly sacrificed technology leadership in order to attain service leadership. The combination of good service and a good basic product usually works better than a poorly serviced, exotic, high-technology product.

FINDING GOOD PRODUCT IDEAS

Studies by Edward Roberts of MIT indicate that most high-technology product ideas for new companies come from positions held with previous employers (source organizations) of the start-up's founders and key employees. Most entrepreneurs get an idea for a product or service based on their current employment.

Other sources of product ideas are customers, customer-sponsored research and development, and product line evolution. The market- and customer-oriented product ideas usually prove to be the best. In fact, at ASK Computer, Sandra Kurtzig was quoted as saying, "Virtually every ASK product evolved from discussions with and suggestions from our customers."

Attendance at technical conferences and trade shows is always a rewarding, if exhausting, exercise, where you will see the latest products being offered.

You should also subscribe to all the trade and business magazines related to your technology area in order to keep aware of new competitive products and trends. Most of these magazines are free controlled-circulation publications. To qualify for higher-level magazines such as *Electronic Business*, you may have to use your imagination in filling out the reader qualification card. In other words, claim that you do recommend, purchase, and authorize purchases for most products and services listed in all the little boxes you checked.

Subscribe to targeted periodicals related to your technology. For example, if your technology or skill is focused on developing programs for the Apple Macintosh computer, you will want to read every issue of *MacWeek*, *MacWorld*, and *MacUser*, as well as technical society journals such as the Institute of Electrical and Electronic Engineers' *Computer, Communications of the Association of Computing Machinery*, etc.

Finally, subscribe to general business periodicals such as *Business Week*, the *Wall Street Journal, Inc.*, and *Success*. *Forbes* may be too *Fortune* 500 oriented, and *Entrepreneur* may be too blue-collar, small-business oriented for your tastes.

Do You Offer the Drill Bits (Means) or the Holes (Ends)?

Apply your technology to develop a product that delivers a benefit to your customers. The most important lesson for an engineer to learn is that one does not sell technology, one sells benefits. There is a saying that "the hardware store owner does not sell drill bits, he sells holes!" You need to first find the holes in your market that need filling with your technology, then you can design and develop your product. Again, your company must be market- and customer-driven and technology-fueled. This is not to say that you need to be marketing-driven. A good product that fills a need and delivers a benefit can be sold with normal marketing effort. You will note that almost all discussion on selecting a product takes us back to marketing principles—not to the state-of-the-art of technology.

Do Not Confuse One Product with a Business

Many engineers have a good idea for a single product. Your market analysis will tell you whether or not you will be able to grow a business based on a single product. In almost every case, though, you will need to identify a family of related products to build a solid business. Always be thinking in terms of a product family and growth potential. This simple observation is often overlooked in the excitement of a start-up.

It is a good idea to have two folders on your desk labeled "next year" and "five years out." Put ideas and thoughts as they come to you in these folders for future analysis. Keep clippings, too, of articles that relate to your technology and marketplace. Brainstorm, and constantly consider new products your company could develop and produce. Today's innovative corporations, such as Minnesota Mining & Manufacturing (3M), derive more than 30% of revenues from products produced in the past five years. Hewlett-Packard gets more than 50\% of its annual revenue from products introduced within the previous three years. Follow the successful models.

MARKETING AND COMPETITIVE ANALYSIS CONSIDERATIONS

Product Positioning and Your Competition

Your temptation to develop a particular product based on your technology fuel must be viewed in the light of product positioning analysis. Figure 10.1, motivated by Francis and Heather Kelly's *What They Really Teach You at the Harvard Business School*, illustrates that a successful product must have a perceived benefit to the customer over other similar products in terms of price and/or performance (as evidenced by feature differentiation).

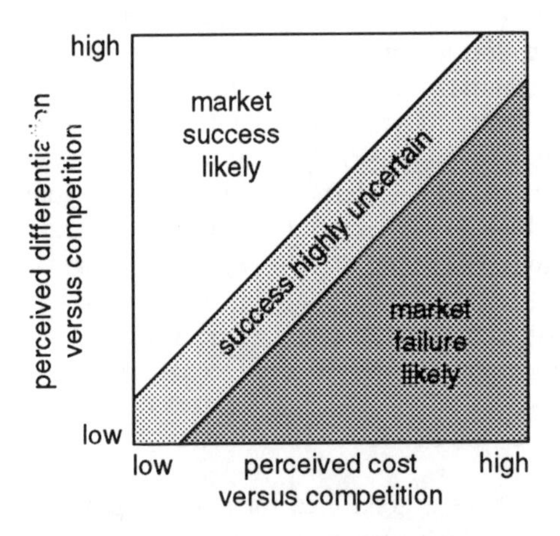

Figure 10.1 Cost versus Perceived Differentiation Model

Competing on the basis of cost alone is clearly an option. Commodity products such as computer memory modules often gain market share primarily on the basis of price. If your technology enables you to produce a needed product at a cost advantage, then that is a reasonable candidate for your start-up's first product. Cost advantage alone will not guarantee you a market presence, however.

Even if you could develop a software program equal in functionality to Microsoft Word or Excel and sell it at half the price, it may not take off.

Customers know the Microsoft brand name, place emphasis on the de facto standard, and perceive a better deal due to better customer service, extended product life, and future product upgrades. Never forget: perception is reality.

Competing based on perceived differentiation of product features and performance is your alternative. Differentiating features can be solely product related and highly visible, for example, adding three-dimensional data representation and charting capabilities to a traditionally two-dimensional spreadsheet program. Or, differentiating features could be less tangible, such as your company's reputation, service, delivery, training, or product support. Figure 10.2 illustrates the positioning of Adam Osborne's low-cost computer programs offered by his Paperback Software International, Ltd. How do you think he did?

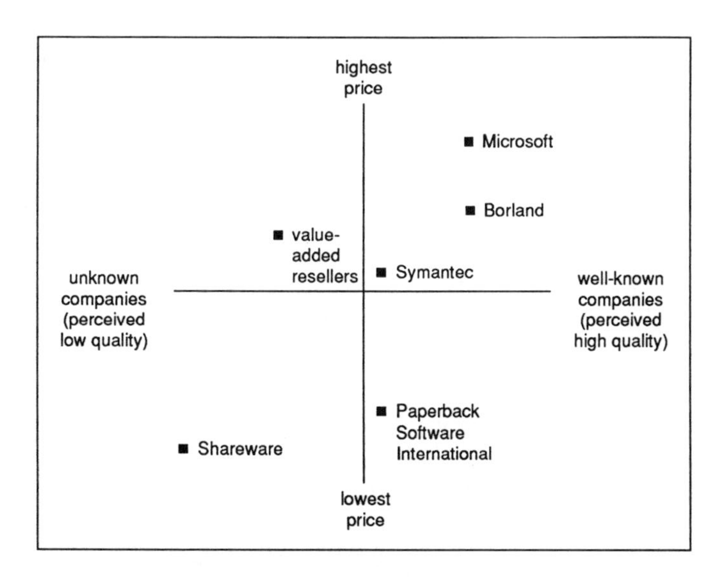

Figure 10.2 Positioning of a Software Product

His product positioning was unique. Paperback's products were clearly differentiated by cost, and their functionality was perceived to be almost as good as that of his more expensive competitors. The company, although new, was well-known because of Osborne's acclaimed reputation (Osborne Computer Corporation offered the world's first portable personal computer with full software for under $2,000) and extensive press publicity. It was destined to be a clear win, except that his competitors sued him for copyright infringement and won. The business did not succeed.

Before you launch a start-up with your product, plot its positioning with regards to your competition and know that there will be sufficient perceived benefits in cost and features.

Maintain a Proprietary Advantage, or Patent Protection

Investors like to hear that their potential entrepreneurs have technical reasons why their product can deliver better benefits to customers and that they can protect those advantages through either trade secrets or patents. Patents and trade secrets will provide entrepreneurs with an opportunity for a head start on their competitors, along with a period of time where they alone can handsomely profit. Many prescription drugs fall into this category. While they are under patent protection, the manufacturer can often charge aggressive prices for those products in heavy demand. Look at your product ideas in this light and try to identify similar opportunities.

Evaluate your proposed product against the competition. One rule of thumb is that a new product should have three to ten times the advantage in terms of price and/or performance of an existing product for a customer to switch. A popular story illustrates this point. Ely Callaway, a 72-year-old former textile tycoon and newcomer to the golf business, catapulted his eight-year-old Callaway Golf Company from a tiny specialty outfit selling mostly novelty clubs into the fastest-growing golf club maker in the country. Callaway explained in a *Business Week* interview:

> You've got to create a product that is demonstrably superior to what's available in significant ways. And—most important—it has to be pleasingly different. That's all there is to it. Simple.

His golf club is more than twice as expensive as a normal driver, but it has a much sought-after larger sweet spot due to its revolutionary design consisting of a larger head with better weight distribution.

However, be careful about claiming design superiority—it can come back to bite you if a competitor later claims you have infringed his patent. The competitor can use your statement to support a claim that he deserves high royalties—you have admitted the design's superiority; obviously, he deserves to be compensated for such a revolutionary and superior idea.

Are you going to develop new word processing software for a personal computer? Your potential customers likely may not consider buying a new, unknown word processor even if it is faster, more functional, and less expensive than what is now being used. The point here is to not become too attached to your technology, invention, or product. Take a good, solid, objective look at what you will produce and the value and benefits it will deliver to the marketplace.

Play in a Large Enough Market

One of the biggest mistakes engineers can make is to try to start a business in which they will develop a specialized product that has a worldwide market potential of only a couple of million dollars. Not only should your target market segment be large enough, it should also be healthy and growing.

Avoid Playing in a Marketing Market

Marketing-driven companies (such as ones selling cigarettes, food products, or perfume) thrive not only on the quality or uniqueness of their products and services but also on their unique methods of promoting and selling their products. This is not the market you should be in.

You may have a technical idea for a new razor blade that will work better than any on the market. To sell this razor blade, however, you would need to compete against giants who spend tens if not hundreds of millions of dollars to introduce such new consumer items. As a technically oriented entrepreneur, you should stick to items that do not compete in the everyday consumer market.

Identify Concrete, Real Customers

If you can list the names and phone numbers of two to five people who, when asked, "Would you buy this product for this price now if it were available from this person?" would respond "Absolutely, yes!" then you are on the right track. Having these customers lined up will do amazing things with regards to investors' reception of your business plan. In fact, some of these potential customers are good candidates themselves for seed-level funding sources.

It is also important that you understand the buying cycle of your customers. Many big ticket items are dearly wanted by customers, but they have to budget for them. This can easily add one year to their purchase action. Also, there is the uncertainty of capital expenditure cycles, which ultimately influence buying behavior in every industry.

Make Your Product Easily and Clearly Understandable

Many engineers concoct elaborate product ideas. That is, when asked "What do you intend to develop or sell?" they need five minutes to answer.

You must be able to describe your product and its benefits with utmost clarity. As one venture capitalist is fond of telling his courting entrepreneurs, "When you tell me what your product does and why someone needs it, I want to hear an answer that's as clear as if you were describing the function of a parking lot; everyone knows what parking lots are used for, and why they are needed." If you cannot describe your product in simple understandable terms, it is unlikely that your potential customers will even know that they need your product.

EXCEPTIONAL PRODUCT ATTRIBUTES

Developable and Producible in a Timely Manner

Heavy investment in plant and equipment, such as that required in semiconductor manufacturing, is risky and expensive. Software, on the other hand, has almost no production costs, yet development and maintenance of the product can sometimes be difficult to manage. Make sure that you do not select a product that will take too much time and money to develop and produce. Again, to yield high returns on investment, you need to attain rapid profitability.

Time to market is critical in this day of time-based competition. You need to be able to produce your first product in a short time to compete. Kurtzig of ASK Computer stated the case well:

> The faster my rudimentary product hit the market, the more money my potential users would save—and, of course, the faster I'd begin earning royalties. I didn't want to fall into the R&D trap of trying to create the perfect mousetrap. The trick is to be in the marketplace just as the demand for it is accelerating. By the time VisiCorp finally came out with their IBM product, VisiOn, Kapor's software, Lotus 1-2-3, owned the market. Adam Osborne, on the other hand, was right on time with the first portable computer. But Osborne then took too long to respond to the trend to IBM compatibility, and within another year his company had gone bankrupt.

High Gross Margins

Software is probably one of the highest gross margin businesses an engineer can start. Because you are small and have limited capital, you need to be able to rapidly produce your first products at low cost and sell them for an aggressive price.

Substantial Collateral Revenue

A good product generates revenue from collateral items such as service and maintenance contracts, accessories, updates, and operating supplies. You want your customers to keep coming back to you.

Clear Distribution Channels

Before making a final decision to sell a particular product, it helps to visualize how that product will ultimately be marketed and sold. For example, if your product will sell for $50,000, it will most likely have to be sold through direct sales representatives. That means hiring and training an expensive staff of sales people who will draw salaries for a long time before sales generate enough contribution margin (gross profit) to cover salaries and commissions for the sales force. A product selling for less than $5,000, on the other hand, will almost always have to be sold through distribution. That is, you would engage either independent distributors (who technically buy the product from you, and take title to the product), or independent sales representatives (who represent your company's products, as well as other competitive products, perhaps). Either way, you will give up typically 40% of the sales price for their efforts. While this seems expensive, being able to identify and use a known, proven distribution channel that will readily accept and promote your product will be very helpful. Proof of an existing distribution channel will also be a selling point with your investors. Many a start-up has produced a great product, only to begin a very long, missionary sales effort that took the company under. Time is money, and trying to establish an in-house sales force or to enlist a reluctant distribution channel can cost you plenty. Produce a product that you know can be readily sold.

PRODUCING YOUR PRODUCT

Assume now that you have zeroed in on a good product idea. You have talked to possible customers and established to your satisfaction that there is a large and growing market where this product can solve real problems. This product could also be the basis for a lucrative product family. Compared to the competition, you seem to be well positioned and protected. The product is easy to describe, and its benefits are obvious. It can probably be developed and produced rapidly, and it has the potential for high gross margins and substantial collateral revenue. Now you have to lay the plans for actually producing your product.

Statement of Requirements

The initial step when devising a development plan for your product will be to give it a very specific description. This can be done by generating a *statement of requirements* document. This document describes the general requirements of your product and is to be used by engineering, marketing, and sales personnel during the product development, testing, and market introduction stages. While this document is primarily a marketing statement, it does cover engineering topics too. Much of this information can be extracted directly from a well-researched and well-written business plan. A good number of start-ups neglect to produce this document, which should not take more than one month to generate if the overall business is well planned. Start-ups lacking a carefully planned product description usually pay the price by heading off in a wrong direction. The statement of requirements gives your development team the direction it needs to make progress without your constant supervision. Figure 10.3 is a sample outline for the statement of requirements document adapted from a successful developer of industrial software.

- What is the product?
  - features
  - applications and uses
  - benefits delivered
  - needs met
- Market analysis and requirements:
  - competition
  - pricing
  - Who is the user?
  - Who makes/influences the purchase?
  - What are likely sales channels?
  - marketing communications and public relations literature
  - distribution channels
- Product requirements:
  - competitive positioning
  - target production costs
  - Who installs the product?
  - training and field support requirements
  - customer support requirements
  - warranty policy
  - upgrade policy
  - user, reference, installation manuals
  - product packaging
  - copy protection policy (for software)
  - maintenance considerations
  - expected product life
  - release schedule (alpha, beta testing, first release, etc.)
  - future product enhancements and extensions
- Functional requirements:
  - performance requirements (responsiveness, accuracy, reliability, mean time between failure, etc.)
  - systems requirements
  - human factors
- External requirements:
  - environmental requirements
  - office, factory, etc.
- Other requirements:
  - regulatory requirements
  - international and export considerations

Figure 10.3 Sample Statement of Requirements Outline

Functional Specification

Following the generation of the statement of requirements, a *functional specification* document should be produced. This is more of an engineering than a marketing document, stating in greater detail requirements relating to the product's function (but not the product's form) such as:

- how fast a machine must process parts, or how fast software must execute instructions or perform mathematical calculations
- how accurate (with what precision) a computation must be
- how repeatable a mechanical part positioning must be
- how a device or program must be controlled by an operator and what operating options must be presented
- how long an instrument must operate before calibration is required
- how heavy a device can be
- how much space a product may occupy

Figure 10.4 is an abbreviated sample outline for a typical functional specification document, again adapted from a successful developer of industrial software.

Terminology
Hardware platform
Operating system
User-interface standards
Help system
Input devices supported
Copy protection
Product features (an itemized feature list related to the application)
Documentation, comment, and annotation facilities
Cross-reference facilities
Debugging support
Fault detection and handling and error recovery
Cut and paste interfaces
Language specifications
Instructions supported
Limitations and restrictions
Program inputs and outputs
Operating environment
Reliability considerations
International language version considerations
Development notes

Figure 10.4 Sample Functional Specification Outline

The functional specification can turn out to be a crucial document. If it is too vague, the customer will use it as a hammer against you—what started out as a bicycle could turn into a Mercedes (both are vehicles with wheels that transport a passenger from point A to point B), and you (the designer) could be required to pay for upgrading it.

A really good functional specification will be broken down into as many sub-specifications as one can reasonably think of. In addition, it will also have three performance target specifications (the data often being listed in three columns) as depicted in Figure 10.5. For this reason, a functional specification is often also referred to as a *performance specification.*

In regard to Level 1 in Figure 10.5, be cautious. This is a danger area for technology companies. The developing company's engineers have been known to get together with the customer's engineers and begin adding bells and whistles ("gee, if we just tweaked this a little bit here, it could do this, too...")—significantly increasing development time without having management pass a long the price increase to the customer.

For Level 3 in Figure 10.5, make sure you look at how much time you add to the development project as well as how much you add to the cost—and price it accordingly. Cost overruns occur because no one did a realistic cost analysis or time analysis, or if they did, the analysis was not updated.

performance level →

function	Level 1	Level 2	Level 3
	what the customer thinks he or she really wants *minus* a few questionable features that would cost quite a bit to develop	what the customer thinks he or she really wants	what the customer thinks he or she really wants *plus* a few extra desirable features that would cost very little extra to develop

Figure 10.5 Performance Specifications

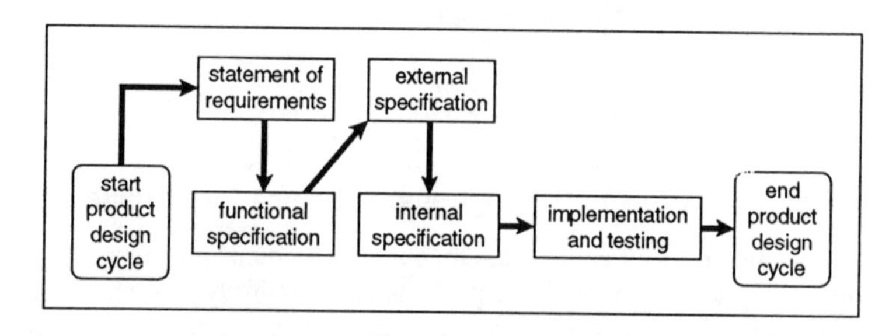

Figure 10.6 Product Design and Development Methodology

As illustrated in Figure 10.6, these documents represent only two of the important stages involved in a good product design and development methodology. Some of your documents will be living documents. That is, they will never be entirely completed and will always be changing. Yet it is important to be diligent in producing and maintaining these vital documents that state what the product is supposed to do, and why. Producing drawings, sketches, and more detailed designs will further propel you toward a closer understanding of the product development task that awaits. Without such documentation, you run the risk of producing what is instead easy, convenient, or interesting to produce as time passes, not what you know your customers want, need, and will buy!

External Specification

You probably do not need to get into this level of planning at business launch time, but you should know that an external specification document is often developed to translate the "whats" in the functional specification into more specific "hows." The external specification states everything about the product that can be seen, felt, measured, or touched by the customer—that is, seen externally.

For example, if a functional specification says to keep the weight under 100 pounds, a corresponding external specification might stipulate that a frame be made of lightweight aluminum. Statements of speed requirements might translate into the types of motors to be used, and statements of accuracy required might translate into the specification of the use of air-guided, frictionless bearings. For software, the external specification might consist of pictures of all menus, displays, dialog boxes, etc., with which a user would interact to perform the function specified.

Internal Specification

This level of specification explains how you will accomplish the external specification. The customer typically does not see or care about this level of detail. For example, in software products, an internal specification might stipulate the language and data structures to be used, along with the names and behavior of major routines, objects, modules, etc. For hardware products, actual component-level decisions are made, down to the part number level if possible.

The internal specification should come last, but unfortunately it is often the first step in a start-up's product design effort. It is very tempting to get started on building it without knowing what the "it" is and what "it" is supposed to do. Writing internal specifications (or worse, building your product prototype) before understanding what benefits the product must deliver by solving which customer problems is exercising poor judgment.

Implementation and Testing

This step should be unambiguous. There should be no more questions about how well the product is to operate, with what parts, and how everything interconnects. If the preceding specifications are implemented properly, the rest should be simple, and major milestones should be met on schedule. Of course, adjustments are constantly required because nothing can be precisely specified in advance and market needs are continually-moving targets. Be sure the best job possible is being done on each level and that the tasks are being undertaken in the proper order. It is amazing how many companies try to start by "hacking" a product without following any design methodology. Most of these enterprises are likely to fail.

If your start-up seems to be working from the bottom up, watch out! You cannot build a product from the pieces until you know what it is supposed to do in some detail. Most inventors are not good product development engineers. If your engineers do not understand the essential role of marketing's input to product specification and do not practice a top-down process of specifying, designing, and developing a product, the business may fail even if the product idea itself is fantastic.

One recent alternative to the traditional top-down method is a rapid prototyping methodology whereby, through iterative experimentation, one converges on a satisfactory design. This method has found some success in today's rapid-time-to-market environment, especially in software projects where one has to discover many of the hows through experimentation. This method is often justified on the basis that many software engineers and programmers are not very manageable anyway, and they tend to produce what is at the moment challenging and of interest to themselves. While there may be some basis for that reasoning, it is still essential that the programmers in this case have a keen sense of what the market needs. Those businesses that have successfully used rapid prototyping follow the rule that designs (the prototypes) are to be reviewed often, and any software written is to be reusable.

Sometimes a *skunk works* is allowed to exist. This is a slang term for an unofficial effort that is allowed to exist more or less out of sight. Unfortunately, out of sight means without much management direction and control. This model might work in some start-up situations where each employee is an exceptional team player, communications are superior, and there is a good sense of the market. If you choose to operate a skunk works, you had better be an active participant.

Chapter 11
WRITE YOUR BUSINESS PLAN

"It's not the writing that's difficult—it's the thinking."
—James Leigh

FORM VERSUS CONTENT

This is the longest chapter in the book. That is not because writing a business plan is complicated in itself, but because the business planning process (content generation) behind the writing (the form) is very difficult. Determining the appropriate form of your plan—which topics to put in, in what order, and with what emphasis—is important. Even more important, however, is that the content of your plan conveys an intimate understanding of what will make your business succeed. Much of what is covered in this chapter involves the content as well as the form of your business plan.

TYPES OF BUSINESS PLANS

First, know that there are two types of business plans: a general *planning and funding* document written to plan for the beginningof the business and to raise funds, and an *operational* business plan to monitor and control the growth of the company. In launching your start-up, you will be working with the former, and therefore that is the one addressed in this chapter.

Other than your last will and testament, the most important document you will ever write is your business plan. A business plan gives birth to your start-up. It enables you and your team to envision and plan how the business will be run and to raise funds. Because both you and your investors have similar questions, one business plan serves both parties. Later, your operational business plans (which can be less comprehensive) will enable you to monitor, plan, and adjust to changes in the growth of your business.

GETTING STARTED

Chapter 6 describes the financial stages of a company's growth. It is important to understand that there are also five operational stages to your company's growth. C. Gordon Bell's *High-Tech Ventures: The Guide for Entrepreneurial Success* clearly describes these well-recognized operational stages, as illustrated in Figure 11.1.

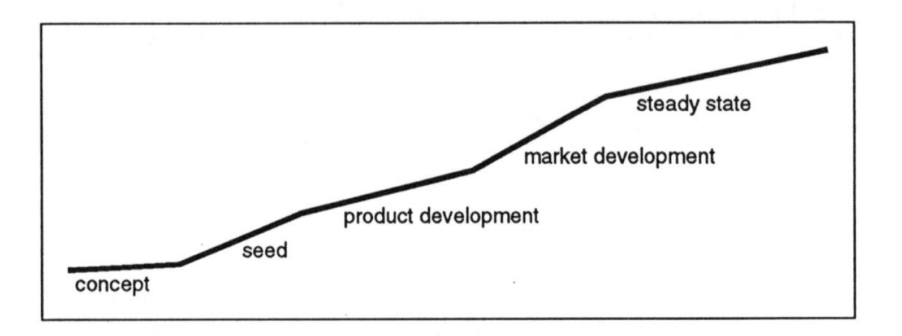

Figure 11.1 Operational Stages of Company Growth

You will be writing your business plan during a concept stage, before you have received outside funding. During a good part of this period you may still be employed.

The *concept* stage is the period during which you and your founding management team develop an idea and write the plan to implement that idea. The financial purpose of the plan is to seek funds for seed-stage testing and refinement of the idea. In some cases, the seed stage can be skipped.

In the *seed* stage your ideas are refined, and an even more detailed business plan may be written for the start-up funding round.

In the *product development* stage your product is developed, tested, and refined before any real sales are made.

The *market development* stage arrives when the product is sold and you beome profitable.

Steady state is reached as the business matures and sustains itself. It should still be growing at this time, and the original investors may decide to exit or cash out.

FIND A TEAM AND WRITE A PLAN, OR WRITE A PLAN AND FIND A TEAM?

You are encouraged to build your founding team as soon as possible. If you cast the vision and establish the mission together, you probably will be more successful. A 25-year study of high-technology companies by MIT professor Edward Roberts shows that multiple founders increase a start-up's chances for success. Larger founding groups start with more capital, generate more sales earlier, and can work longer hours. In one subgroup of 20 young companies, 63% of those with more than two founders performed better than average, while only 20% of those with one or two founders exceeded average performance.

Not only will you need help to write a good plan, you will need help in raising funds. Therefore it is a good idea to team with partners who can assist you in attracting funds. Investors do not throw money at glossy business plans; they invest in teams with plans to exploit market opportunities. This subject is also discussed in Chapter 8, which deals more extensively with creating winning management teams.

WHEN TO WRITE THE PLAN

When do you write the business plan, and with whom? If you are currently employed, can you take the time and energy to write your business plan and not have a conflict of interest or commitment with your employer? During the plan-writing phase you will need to conduct extensive research, talk to potential customers and investors, and understand your market better. Rarely can all this be done successfully on a part-time basis.

Silicon Valley trade secrets and unfair competition lawyer James Pooley addresses the legal implications of when you must leave an employer:

> You may find that planning your move while you're still working for the company is somehow unethical. Deliberate, careful planning of your new enterprise is neither illegal nor immoral. Unless you are independently wealthy, you must do it while you still have an income and before you burn any bridges. As long as you don't actually begin your competing business or start recruiting your team from your employer's staff while still on the payroll, you should be clear.

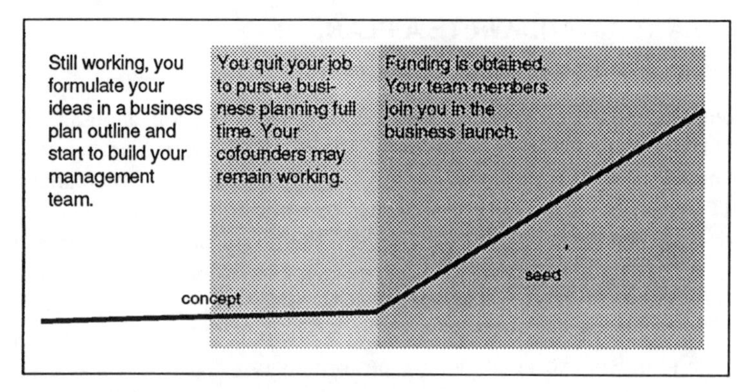

Figure 11.2 Start-Up Scenario

The start-up scenario illustrated in Figure 11.2 shows that at some point in time you must quit your job and dedicate yourself full time to engineering your start-up. It is for this reason that there are thousands of voyeurs and would-be entrepreneurs who will never start that business they constantly dream of. It indeed takes quite a sacrifice and risk to start your own business, especially for the first time. For this reason, many start-up founders are people who have been fired or come from other failed start-ups. In fact, many people claim that the purpose of starting your first business should be to really achieve success in your second. Viewed from this perspective, if you are not yet sufficiently committed to quit your job and launch your dream, maybe you should consider joining someone else's start-up as a key employee or cofounder. That way you would only need to quit your job after seed funding has been obtained, and you would know where your next paycheck is coming from.

HOW LONG SHOULD IT TAKE TO WRITE YOUR PLAN?

You should plan to spend three to twelve full-time months in the unfunded concept stage of your business while you write your business plan and launch your start-up. Michael O'Donnell says, "A typical time frame for writing a plan would exceed 300 hours over one to six months, depending on how much time can be devoted to the research and writing." Ronstadt and Shuman state that "[in] our experience [the time required to write a business plan] ranges from 100 hours to nearly 2000 hours." Your founding team members, as you convince them to join your venture, will of course help you to write the plan. As the CEO, however, the burden is squarely on your shoulders to take the risk and ensure that the plan gets written.

ESSENTIAL TOOLS

Almost everyone, especially an engineer or a scientist, has access to a personal computer today. If you do not own a computer and good word processing and spreadsheet software, you must go shopping. These are essential tools of the trade these days. Invest in (or at least obtain easy access to) a good printer also, preferably a laser or ink-jet printer rendering 300 dots-per-inch resolution. Your plan should reflect both good form and good content. Ignore the form and you will lose many potential investors. You know it makes good business sense to dress for success, so dress your business plan as well; it is an extension and reflection of yourself. Like a resume, it introduces you and says, "This is the very best I can do; now do you want to talk to me further?"

A sidebar to this chapter reviews some excellent commercially available word processing and spreadsheet templates to aid in organizing your business plan that you will definitely want to look at.

GOOD BUSINESS PLANNING

There are dozens of good business plan models and actual business plan samples to be found in literature. If you have access, try to rummage through other private business plans to find examples that are especially relevant to your start-up situation. By now you should be networking with potential team members who have access to a variety of business plans.

Rummaging through other private business plans to which you have access can be dangerous if it results in illegal misappropriation of another's trade secrets. Watch out for violating confidentiality agreements when you do this.

The following are business plan basics and some good sample outlines.

BUSINESS PLAN BASICS

This book is built around six elements of success of a start-up business:

1. management teams
2. markets and customers
3. products
4. plans
5. funding
6. luck and persistence

These success elements map into the four substantive sections of any business plan:

1. management team
2. marketing
3. products
4. financial projections

Your business plan must clearly cover these critical sections, and there are several variations in format. As you choose a plan format, never lose track of the following basic messages you need to give:

- an executive summary compelling the reader to study the plan
- a management team that will guarantee success in the venture
- a market opportunity that gives this business a distinct competitive advantage. You should state who will buy your product and whether it is in a new or existing market. (Chapter 9 contains more extensive material that should be included in your plan.)
- a product that is producible and merchantable (state whether this is a new or existing product)
- financial projections that satisfy return-on-investment objectives, are achievable and believable, and obviously do not violate rule-of-thumb sanity-check ratios

PLAN EMPHASIS

Emphasis in a business plan is usually put on either the management team or the growing market opportunity. Decide what you are offering to your investors, and structure your plan accordingly. If your management team is inexperienced and this is your first business launch attempt, you must focus on your unique market opportunity and how your high-technology product will make the business successful. Sometimes, personnel resumes are buried in a plan. Since many investors want to read about the management team first, if this material is hard to find, your plan may not go very far. If you have a successful management track record, on the other hand, or your combined management team looks very appealing, definitely highlight this asset as the first and most visible section of the plan by placing it immediately after the executive summary.

Different investors have differing postures on the relative importance of a management team versus a growing market opportunity. You will often hear statements such as, "we invest in people, not ideas; we assume the product idea can be built," "the market is number one," or "lack of a management team is the number one company killer."

If your start-up has a strong management team and if you are playing in an attractive, growing market, then the Genms example illustrated in one of this chapter's sidebars should suit you well. You can make your plan short and compelling, especially if you have a strong story to tell. Like a powerful person speaking in a soft voice, it forces everyone to listen attentively. A plan that is too comprehensive, discussing every operations detail, can come off like a not-so-powerful person who is trying too hard. You are selling yourself, your team, and your unique market opportunity, not a used car. Prepare and present your plan accordingly.

There exist dozens of comprehensive plan outlines for the general planning and funding type of business plan. Three of these outlines are presented in this chapter. The detailed, lengthy, comprehensive plans are not recommended. If nothing else, the items your investors will be looking for (people, financial projections, and unique market opportunity) will be too difficult to find amid a long plan. However, you can use these more detailed plan outlines as checklists to see that nothing important has been missed.

NEW VENTURE BUSINESS PLAN OUTLINES

JIAN's BizPlan*Builder*™ suggests the following business plan outline:

A. Executive summary
B. Present situation
C. Objectives
D. Management
E. Product/service description
F. Market analysis
    1. Customers
    2. Competition
    3. Focus group research
    4. Risk
G. Marketing strategy
    1. Pricing and profitability
    2. Selling tactics
    3. Distribution
    4. Advertising and promotion
    5. Public relations
    6. Business relationships
H. Manufacturing
I. Financial projections
    1. 12-month budget
    2. 5-year income (profit and loss) statement

   3. Cash-flow projection
   4. Pro forma balance sheet
   5. Break-even analysis
   6. Sources and uses of funds summary
   7. Start-up requirements
   8. Use of funding proceeds

J. Conclusions and summary

K. Appendix

Adapted from JIAN BizPlan*Builder*™, with permission.

INSTITUTIONAL VENTURE PARTNERS' SUGGESTED BUSINESS PLAN CONTENTS

The following business plan outline was recommended by Norman A. Fogelsong and Kenneth J. Kelley from Institutional Venture Partners. IVP is a leading venture capital firm in Silicon Valley specializing in investments in start-up and early stage high-technology companies.

A. Executive summary with 5-year milestones

B. Product or service description

C. Business strategy overview

D. Marketing and sales plan
   1. Market size, projected growth, and segmentation
   2. Competitors and their market shares
   3. Strategic positioning and marketing plans
   4. Channels of distribution
   5. Sales strategy and 5-year sales forecast
   6. Customer references or references on market potential

E. Operations plan
   1. Development and engineering programs
   2. Manufacturing and materials programs
   3. Facilities plan
   4. Product service or maintenance programs

F. Management and key personnel
   1. Organization
   2. Detailed resumes with personal references
   3. Staffing plan
   4. Stock option plan or incentive program

G. Financial statements and projections
   1. Historical and current financial statements
   2. Annual projections of income statement, balance sheet, and cash flow for next 5 years

3. Monthly projections of income statement, balance sheet, and cash flow for next 1 or 2 years
4. Existing shareholders and ownership percentages

H. Proposed financing
1. Amount and terms
2. Post-financing capital structure
3. Use of proceeds

I. Appendices

In a recent address to a Silicon Valley engineering management society meeting, Fogelsong and Kelley of IVP gave the following hints for engineers wanting to start their own companies. First they said that an attractive business plan will exhibit:

- a talented management team with entrepreneurial skills
- a large market need with high growth potential (we are looking for a $50 million to $100 million company valuation in five years)
- a unique and/or proprietary technology
- sound and executable plans
- attractive financial returns (we are looking for 10 times returns over five years)

They also claim that the keys to your success will be:

- Vision: You need a clear vision of what the company should be.
- Commitment: You need the commitment to work hard to succeed.
- Focus: You need to focus on the task at hand.
- Execution: You need superb execution of your plan.

A PERSONAL FAVORITE

The following combined business plan outline and checklist has been used in a number of successful high-technology engineering start-up opportunities. Again, do not attempt to cover every item listed; use this outline to verify that you have not missed any important topics for your business situation.

A. Executive summary
1. Objective of the business
2. Background and unique opportunity and market
3. Management team

4. Products
    a. Initial
    b. Future
5. Marketing strategy
6. Producing the product
7. Investment sought and return
8. Use of proceeds

B. The company's objectives
1. Origin of business idea and mission
2. Current status; immediate and long-term objectives
3. Meeting a new/existing market's needs
4. Raising seed money
    a. Mission statement (must be clear, precise, and compelling)
    b. Approach for initiating new organization
    c. Time schedule for starting the business
    d. The unique opportunity
    e. Description of business objectives in clear, simple, nontechnical terms
    f. Long-range objectives
    g. Short-range goals
    h. Character and image of business

C. Background and the market opportunity
1. The competition
2. A growing market
    a. Meeting competition
    b. Market growth data
    c. Competition (include printed material in appendix)
    d. Who buys product now, for what, where, and when?
    e. Needs and wants of intended market segment
    f. Market survey data used to develop plan and selecy market niche

D. Personnel
1. Qualifications to run the business individually and as a team
2. Organization of the business at present
3. Organization of the business after funding
4. President and CEO
5. Marketing manager
6. Treasurer
7. Secretary
8. Engineering management

9. Board of directors
    a. Resumes of management team and qualifications track record (select a consistent, powerful format)
    b. Organization chart (pre-funding, post-funding) of officers, board of directors, key employees

E. Product description
1. First product
2. Design drawing (attached; investors love to see pictures, sketches, and drawings in business plans)
3. Functional specifications
4. Sample (mock-up is okay) advertising brochure including sketch or photo
    a. Description of product with drawings, sketches, pictures, or illustrations
    b. Desirability, advantages of product
    c. Present state of the art, trends, predictions for your place or niche
    d. Patentability or uniqueness of product (Do not include too much technical information in your business plan. Specifically, technology-related details that you consider trade secrets probably should not be written down in your business plan. Business plans frequently circulate far beyond your wildest dreams.)
    e. Describe a family of (future) products (one product usually does not make a business)

F. Marketing strategy
1. Distribution arrangements
2. Direct sales, rentals
3. Sales channels, costs, and calls per salesperson per year
4. Unique promotional concepts
5. Delivery, field support, and maintenance emphasized
    a. Marketing approach (market segment and distribution channels)
    b. Basic selling approach (lead generation, cost per lead and cost per sale, lead generation time, and sales cycle time)
    c. Market share expected over time (By the way, overestimating this number has discredited many financial projections.)
    d. List of three to five people who will buy your new product (Get their names and permission to call them; some investors rely heavily on

potential customer testimony for their due diligence.)

        e. Pricing

G. Development and operations plan
    1. Personnel staffing requirements
    2. Facilities and equipment required to develop and produce product
    3. Make versus buy strategy
    4. In-house production and subcontracting
    5. Research and development required (Keep this to a minimum: investors want a product fast! However, if time plans are not realistic, problems will quickly develop as milestones are missed.)
    6. Operations and manufacturing considerations unique to the product

H. Financial pro formas
    1. Assumptions (State these clearly and explicitly. Format your financial templates so you can easily change any assumption and have the effects immediately filter throughout all spreadsheets. Spend time to set up your spreadsheets right; otherwise minor changes will be hard to accommodate later.)
    2. Profit and loss for 5 years
        a. By quarter, years 1-2
        b. By year, years 1-5
    3. Cash flow and sources and uses of cash for 5 years
        a. By quarter, years 1-2
        b. By year, years 1-5
    4. Balance sheet for 5 years
        a. By quarter, years 1-2
        b. By year, years 1-5
        c. Financials (State assumptions; compute return on investment for your sanity check but do not include in plan. This is explained later in this chapter.)
        d. Profit and loss statements (sales and profits)
        e. Balance sheets
        f. Cash-flow analysis
        g. Break-even chart for minimum sales goal
        h. Fixed asset acquisition schedule by month (item and amount)
        i. Ownership interest reserved for founders
        j. Capital needed
        k. Founders' share of initial capital investment

I. Capitalization plan
   1. History, funding plan, capitalization, and current ownership
   2. Authorized, outstanding and reserved stock, warrants, and options; certain loans and other financial transactions
   3. Use of proceeds

J. Summary and Conclusions
   1. Unique opportunity
   2. High-risk, high-reward investment
   3. Investor qualifications
   4. Time schedule for funding

K. Appendix
   1. Articles from trade journals
   2. Competitors' literature
   3. Resumes (if not included in body)
   4. Product design drawings, photos
   5. Sample brochure (dummy okay)
   6. Customer references

ADDING OR HIGHLIGHTING SECTIONS

It is important for you to take the time to work up an outline before you write your plan and to customize that outline to fit to your situation.

If you attempt to include all of the topics suggested thus far in this chapter, you will have a plan that is much too long and too difficult to assimilate. Know what aspects are important to your business' success and document those. For example, if your start-up will be capital-intensive, a capital equipment acquisition schedule and cost sheet would be exceedingly useful in addition to the standard financial pro formas.

One start-up business had a customer list of five *Fortune* 100 companies from earlier contract work and added a section to its business plan entitled "Existing Customer Base." It made quite an impression on investors. Here were good references that could be checked during due diligence, along with important evidence that this start-up had a head start. Everyone looks for such an advantage. If you make use of customer lists, make sure you do not violate noncompete agreements or misappropriate trade secrets.

Another company was modeled after others in a very successful industry (although they were not to be direct competitors). It added a section entitled "Why We are a Good Investment," which tabulated sales levels and company valuations over a period of time for these model companies.

Investors clearly associated with these success models, and it made the start-up's pitch more credible.

A third company had not only a strong management team, but an active and capable board of directors who truly helped to manage the business. It had a section in its plan entitled "Management and Board of Directors."

Are your product development plans especially complex, or are you proposing to develop a complex process? If so, you will want to convince your investors that you can get there in a reasonable time frame since rapid profitability is the key to high return on investment. You could consider adding a section on schedules and milestones.

Identify the strengths and selling points of your business proposal and promote them in your business plan. Identify any perceived weaknesses in your venture and adequately defend them. Make sure this important material is highlighted in your executive summary and is easy to find in the body of the plan. Use existing guides to assist you and to help you check for completeness, but do not let them bind you.

CLASSIC PROBLEMS

Your new venture could fail for several reasons. In preparing your business plan be on the lookout for the following classic problems:oυ.

- inadequate market knowledge
- ineffective marketing or sales approach
- inadequate awareness of competitive pressures
- potentially faulty product performance
- rapid product obsolescence
- poor timing for the start of a new business
- undercapitalization due to unforeseen operating expenses, excessive investment in fixed assets like buildings or land, or other financial difficulties

USE STANDARD RATIOS

Use standard and expected ratios in your financial cost estimates. For example, if the expenses for marketing and sales in your industry typically run 16—17%, you will have much explaining to do if you show less than 10% or more than 20%. Sophisticated investors examining your financial projections will take you for an amateur if you are too far off on common expense ratios. The same goes for revenues and gross and net profit margins. Study the trade and financial literature in your industry to find numerous ratio examples such as the following.

Marketing, sales, and general and administrative costs as a percent of total revenue:

- HP, a large, stable company, spends about 25% of total revenues for marketing, sales, and general administrative costs.
- DEC's selling and administrative costs in 1989 were 31% of revenues; for Sun Microsystems, the figure was only about 24%.
- SG&A (selling, general, and administrative) is normally high in a technology-based company: about 35—40% of product revenue.
- In the computer business it is common to spend more than 20% of revenues on direct sales, service, and post-sale support.
- High-technology companies often spend 10—20% of revenues on direct sales.

Gross and net margins as a percent of total revenue:

- HP showed a gross margin (total revenues minus cost of goods sold) of approximately 50%.
- A net margin (after-tax profit) of 10% is often considered substantial.
- Net margins (after tax) should usually exceed 10% for a software company.

Research and development (R&D) expenses as a percent of total revenue:

- HP spends about 8% on R&D.
- R&D is normally 15—20% of product revenue in a technology company.

Revenue ratios:

- Revenue from maintenance for a software company might equal 10—18% of the product cost and be subscribed to by 50% of all customers.

Advertising and promotion:

- In technology-oriented companies, a large sum will be spent on documentation, while only a small amount will be spent on advertising and promotion. Most marketing expenses are earmarked for direct sales, training and supporting distribution channels, customer education and application support, service, and post-sale support.

- In consumer businesses, promotional costs can exceed 10% of revenues.

Sales per employee:

- Typically, a company in the expansion stage and beyond will have \$150,000 to \$200,000 in revenue per employee. Revenue of \$300,000 per employee is not unheard of in software companies.

This last ratio of revenue (sales) per employee is very important. You need to obtain industry figures for your business to make sure your plan makes sense. For example, Figure 11.3 (reported in *Electronic Business*, July 22, 1991) clearly shows not only that revenue per employee is high in the electronics industry, but that it is climbing even higher.

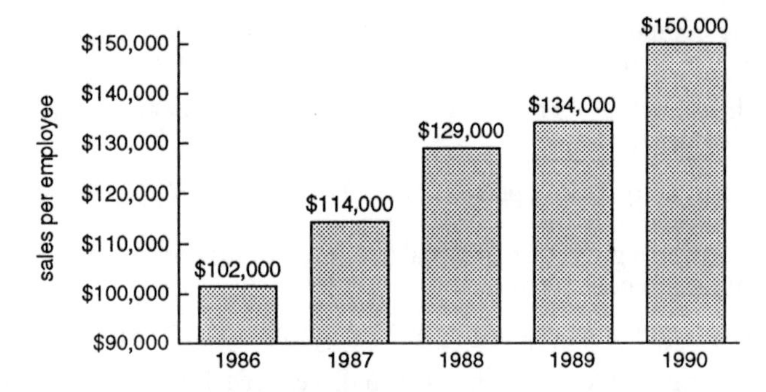

Figure 11.3 Productivity for the Top 200 Electronics Companies

If your business falls in the area of electronics and your plan shows revenues of only \$100,000 per employee, you may not be taken very seriously.

Are you projecting revenues or profit margins above or below investor expectations? Miss these marks and your plan could be quickly dismissed.

It is not difficult to obtain information on these ratios. Visit the library and obtain copies of other business plans from your associates and team members. You should also evaluate several annual reports of young, growing public companies in your industry. You can obtain these reports from any public companies just by calling them up and asking for the investor relations department. Tell them you are a private investor and wish to receive copies of the latest annual report and any subsequent quarterly reports. Ask for their 10K and 10Q, which are more complete versions of these reports that they are required to submit to the SEC and make available to anyone who asks for copies.

See Table 11-1 for some excellent benchmarks if you plan to launch a software-based business. These typical operating expenses as a percent of revenue for software companies come from the September 1991 issue of David Bowen's *Software Success* survey, where he compiled responses from 50 of his subscribers. There are very few financial benchmarks of this quality available for small software companies, and this information can be critical to the successful management of your business.

Table 11.1 Operating Expenses for Software Companies as a Percent of Revenue

Category	1991 (%)	1990 (%)
revenue		
product revenue	76.8	84.0
support revenue	15.4	8.0
professional services	7.8	8.0
	100.0	100.0
cost of sales		
hardware	5.9	3.0
royalties	4.4	4.0
other cost of goods sold	5.5	7.0
total cost of goods sold	15.8	14.0
gross margin	84.2	86.0
marketing expenses		
salaries	3.0	6.0
promotion/lead generation	11.1	11.0
	14.1	17.0
sales expenses		
salaries	5.6	8.0
commissions	4.5	2.0
other	0.6	0.0
	10.7	10.0
support expenses		
salaries	8.6	8.0
other	1.0	1.0
	9.6	9.0
development expenses		
salaries	15.0	13.0
computers	3.0	3.0
other	1.7	1.0
	19.7	17.0
administrative expenses		
salaries	6.8	9.0
facilities	5.3	4.0
other	7.0	8.0
	19.1	21.0
total expenses	89.0	88.0
profit before tax	11.0	12.0

LOOKING LIKE AN AMATEUR

It is difficult not to look like an amateur if you admittedly are one. Most engineers attempting their first start-up will not have had the operations and business experience needed to prepare a first-class business plan, attract capital, and pull off the execution. This is why it is important to build a strong management team to help you launch your start-up. As discussed in Chapter 2, you might even want one of your team members to be the CEO while you take on the position of engineering vice president.

FINANCIAL PRO FORMA DOCUMENTS TO GENERATE AND MASTER

Balance sheet, profit and loss (P&L) statement, and sources and uses of cash (and cash flow) are the three basic financial statements that must be clearly understood by the start-up entrepreneur. If you rely on someone else to generate these statements for you, you will not understand them and you will not be able to defend them. Worse, you will not be able to use them to control your business. If you wonder why a fresh MBA gets paid a lot of money to run a company while engineers seem to be relatively underpaid and have to struggle to get promoted, the MBA's understanding of financial controls is part of the answer. While it takes no genius to get an MBA degree (most engineering degrees are harder to earn than most MBAs), the MBA often has the financial edge. It is suggested that you attend several night school college classes in finance and accounting before you try to start your own company. Your investors will be looking for someone with financial control skills to run your business. Even if one of your team is financially experienced, all founders will benefit from familiarity with financial jargon and procedures.

WORKING BACKWARD

As you start to prepare the financial projections for your business, it is tempting to work backward from the answers. For example, you may have figured out that in five years you need to have revenues of $50 million and 10% post-tax profits in order to allow your investors to obtain a compounded 25% return on investment. If you iterate through a spreadsheet to calculate the required sales price of your product to meet these goals, all you have is a number that works in the model, not in the market. It is okay to calculate this number as a sanity check, just do not use it in your model.

GET GOOD DATA

The hardest part of preparing financial projections is not in the mechanics, but in getting good data. "Garbage-in, garbage-out" is an old computer programming expression that doubly applies here. In the preceding example, you must determine your sales price from the market. What people will pay for your product is based on what benefits it delivers to them and what alternative products cost. You must clearly understand your market and competition before you can create your financial plans. Engineers in particular tend to ignore the marketing side of a business, figuring that sales problems can be dealt with later.

DO NOT IGNORE YOUR OWN DATA

After you know what the market will pay for your product, you can look at your cost-of-goods and backward-computed required price to see if you have a viable business plan. If you do not, you must seriously rethink whether your business is viable or not. Keep in mind that the most important purpose of the business plan is to convince yourself that this is a reasonable thing to do; selling your ideas to investors is secondary. Start a business for which your numbers initially tell you that you will fail and you will probably fail. Many failed start-ups likely had initial business plans that indicated a flaw which was ignored.

Writing a business plan forces you to consider every aspect of your proposed business. Any information you put into the business plan should confirm and reinforce information you earlier relied upon. Resolve any discrepancies to your satisfaction before proceeding.

BALANCE SHEET

The balance sheet financial report is sometimes called a *statement of condition* or *statement of financial position* because it represents the state of the business at a point in time. This snapshot of your business differs from a profit and loss statement, for example, which summarizes activity over a period of time. A balance sheet shows the status of your company's assets, liabilities, and owners' equity on a given date, usually the close of a month or year. One way of looking at your business is as a mass of capital (assets) arrayed against the sources of capital (liabilities and equity). Assets must always equal liabilities plus equity.

Your balance sheet lists the items that make up the two sides of this equation. To efficiently analyze a balance sheet, you need to compare it to prior balance sheets (e.g., to see an improvement or degradation in various positions and ratios) and other operating statements (e.g., profit and loss, sources and uses of cash, etc.).

The sample balance sheet shown in Figure 11.4 indicates the level of detail you should strive for. The actual numbers are relatively meaningless, and are included only as examples. Your business needs will dictate what values should be inserted. Lines with zero amounts should be included so your reader will not think you missed something. In addition to projected balance sheets for the end of each of your first five years, you will also want to prepare quarterly sheets for the first two years. Some investors will insist that you prepare these quarterly balance sheets. Again, your worksheets should be assumption-based, allowing you to quickly enter different estimates to immediately see the results as they filter throughout all your linked financial pro formas.

	Year 1	Year 2	Year 3	Year 4	Year 5
Assets					
Current Assets					
cash	$10,000	$10,000	$10,000	$10,000	$10,000
investments	$1,800	$2,700	$4,050	$6,075	$9,112
accounts receivable	$29,500	$47,200	$75,520	$120,832	$193,331
notes receivable	$5,000	$5,000	$5,000	$5,000	$5,000
inventory	$45,000	$60,750	$82,013	$110,717	$149,468
total current assets	$91,300	$125,650	$176,583	$252,624	$366,911
Plant and Equipment					
building	$175,000	$175,000	$175,000	$175,000	$175,000
office equipment	$62,000	$62,000	$62,000	$62,000	$62,000
leasehold improvements	$18,500	$18,500	$18,500	$18,500	$18,500
less accumulated depreciation	$0	$0	$0	$0	$0
net property and equipment	$255,500	$255,500	$255,500	$255,500	$255,500
Other Assets	$0	$0	$0	$0	$0
Total Assets	$346,800	$381,150	$432,083	$508,124	$622,411
Liabilities and Owner Equity					
Current Liabilities					
short-term debt	$13,500	$13,500	$13,500	$13,500	$13,500
accounts payable	$22,500	$29,250	$38,025	$49,433	$64,262
income taxes payable	$1,500	$1,500	$1,500	$1,500	$1,500
accrued liabilities	$0	$0	$0	$0	$0
total current liabilities	$37,500	$44,250	$53,025	$64,433	$79,262
Long-Term Debt	$22,000	$19,800	$17,820	$16,038	$14,434
Owner/Stockholder Equity					
common stock	$250,000	$272,340	$307,526	$363,199	$451,370
retained earnings	$37,300	$44,760	$53,712	$64,454	$77,345
Total Liabilities and Owner Equity	$346,800	$381,150	$432,083	$508,124	$622,411
Ratios					
current ratio = (total current assets/total current liabilities)	2.43	2.84	3.33	3.92	4.63
quick ratio = (cash + accounts receivable + notes receivable/total current liabilities)	1.19	1.41	1.71	2.11	2.63
return on assets = total assets/net income					

Figure 11.4 Balance Sheet

PROFIT AND LOSS (P&L) STATEMENT

The P&L statement is a summary of the revenues (sales), costs, and expenses of your company during an accounting period. The P&L is sometimes called an *income statement, operating statement, statement of profit and loss,* or *income and expense statement.* Look at the annual reports from a variety of public companies until you become comfortable with the various names and formats for the P&L. It is the easiest of the three statements to comprehend.

The sample P&L (income) statement in Figure 11.5 shows what you should strive for. While this figure illustrates a less detailed profit and loss statement by year for five years, you will also want to prepare a more detailed statement by month and quarter for the first year and by quarter for the second year.

	Year 1	Year 2	Year 3	Year 4	Year 5
Sales					
product or service A	$58,000	$95,700	$157,905	$260,543	$429,896
percent of total sales	58%	61%	66%	69%	72%
product or service B	$22,000	$28,600	$37,180	$48,334	$62,834
percent of total sales	22%	18%	15%	13%	11%
product or service C	$20,000	$32,000	$45,000	$67,500	$101,250
percent of total sales	20%	20%	19%	18%	17%
total sales	$100,000	$156,300	$240,085	$376,377	$593,980
Cost of Sales					
materials	$21,500	$33,325	$51,654	$80,063	$124,098
percent of total sales	22%	21%	22%	21%	21%
labor	$31,000	$44,950	$65,178	$94,507	$137,036
percent of total sales	31%	29%	27%	25%	23%
overhead	$18,500	$24,050	$31,265	$40,645	$52,838
percent of total sales	19%	15%	13%	11%	9%
total cost of sales	$71,000	$102,325	$148,097	$215,215	$313,972
Gross Profit	$29,000	$53,975	$91,988	$161,162	$280,008
gross margin	29%	35%	38%	43%	47%
Operating Expenses					
selling costs	$2,000	$2,500	$3,125	$3,906	$4,883
percent of total sales	2%	2%	1%	1%	1%
research and development	$2,800	$3,780	$5,103	$6,889	$9,300
percent of total sales	3%	2%	2%	2%	2%
general and administrative	$4,100	$4,102	$4,103	$4,105	$4106
percent of total sales	4%	3%	2%	1%	1%
total operating expenses	$8,900	$10,382	$12,331	$14,900	$18,289
percent of total sales	9%	7%	5%	4%	3%
income from operations	$20,100	$43,593	$79,657	$146,262	$261,719
percent of total sales	20%	28%	33%	39%	44%
interest income (expense)	($5,000)	($4,999)	($4,998)	($4,997)	($4,997)
income before taxes	$15,100	$38,594	$74,659	$141,265	$256,722
taxes on income	$5,889	$15,052	$29,117	$55,093	$100,122
Net Income	$9,211	$23,542	$45,542	$86,172	$156,600
percent of total sales	9%	15%	19%	23%	26%

Figure 11.5 Profit and Loss (Income) Statement

Adapted from JIAN BizPlan*Builder*™ with permission.

Notice how much easier your tabular data is to comprehend when you make it into a graph as shown in Figure 11.6. Good business plans should be highly pictorial and easy to read.

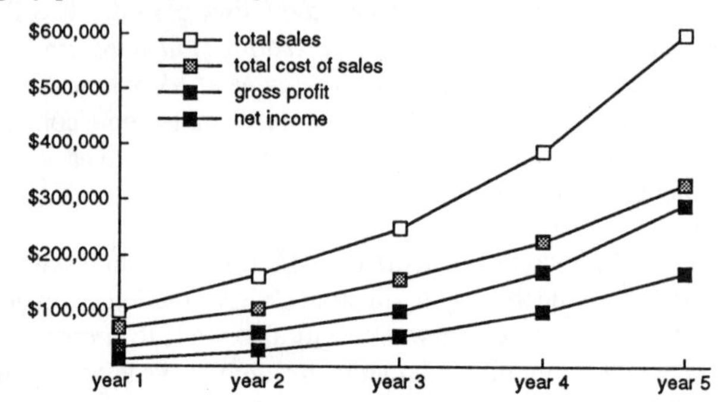

Figure 11.6 Plot of Profit and Loss (Income) Statement

REVENUE VERSUS INCOME

Many people are confused by the terms *revenue* and *income*. In the business situation, revenue denotes the gross figure, namely the sales from the business. Income denotes the net figure: the sum remaining after expenses or cost-of-goods-sold have been deducted (namely the profit). However, there are many different accounting, financial, and common definitions of these terms, and gross income can certainly mean the same thing as revenue in some contexts. Do not be embarrassed if you find yourself confusing the two. Perhaps because salary is often informally equated with personal income, many first-time entrepreneurs correspondingly confuse what one might call a business' sales income (i.e., its revenue), with the business' income. If you study the following equations, the confusion quickly dissipates. In a simple salaried situation, you have no deductible costs associated with producing the services for which you are paid. Thus, your revenue for working is equivalent to your income.

General Terminology	Salaried Terminology	Business Terminology
gross revenue	salary or personal income	sales income or revenue
less cost of providing goods or services	less cost-of-services-provided ($0)	less cost-of-goods-sold (COGS) ($ > 0)
equals income	still equals income	equals profit or income

SOURCES AND USES OF CASH (AND CASH FLOW)

This is the most difficult of the three financial statements to understand. While it is relatively simple to generate and understand a cash flow statement, the more modern sources and uses of cash statement, which includes cash flow, requires a deeper understanding of finance. You can master the sources and uses of cash statement with practice and disciplined study of each line item.

Sometimes called the *sources and uses of funds statement* or *source and applications of funds statement*, this statement allows you to analyze the changes in the financial position (represented by the balance sheet and the P&L statement) of your business from one accounting period to another. You will find that these statements are now required in the annual reports of all public companies, and it is suggested that you study several of them.

To understand the sources and uses of cash statement, you first need to know what working capital is. Working capital is equal to current assets minus current liabilities, and it finances the cash conversion cycle of your business. The cash conversion cycle includes the time required to convert raw materials into finished goods, finished goods into sales, and accounts receivable into cash.

If you are unfamiliar with any of the preceding terms or with concepts such as depreciation, net income, etc., you should pick up a pocket dictionary of finance and investment terms. The sources and uses of cash statement has two parts: the sources of funds summarizes the transactions that increase working capital such as net income, depreciation, the issue of bonds, sale of stock, or an increase in deferred taxes; the uses of funds or applications of funds summarizes the way funds are used, such as for the purchase or improvement of plant and equipment, the payment of dividends, the repayment of long-term debt, or the redemption or repurchase of shares.

The sample sources and uses of cash statement shown in Figure 11.7 again indicates what you should strive for. While illustrated here is a less detailed yearly statement for just the first three years, you would prepare a sources and uses statement by month or quarter for your first two years, as well as yearly statements for five years.

	Year 1	Year 2	Year 3
Source of Funds			
income after taxes	$54,500	$163,500	$490,500
depreciation and amortization	$22,000	$22,000	$22,000
operating cash flow	$76,500	$185,500	$512,500
increased long-term debt	$40,000	$40,000	$40,000
issuance of stock	$100,000	$50,000	$100,000
total source of funds	$216,500	$275,500	$652,500
Use of Funds			
marketing and advertising	$25,000	$37,500	$88,000
salaries	$15,000	$15,000	$60,000
facilities	$65,000	$70,000	$90,000
capital equipment	$18,000	$22,500	$101,000
research and development	$20,000	$40,000	$85,000
operations expenses	$22,500	$23,500	$66,000
cash dividends	$0	$0	$0
increased working capital	$51,000	$67,000	$162,500
total use of funds	$216,500	$275,500	$652,500
Summary of Changes in Working Capital			
decreased cash	($20,000)	($20,000)	($47,000)
increased accounts receivable	$32,000	$30,000	$181,000
increased inventory	$40,000	$58,000	$52,500
increased accounts payable	($16,000)	($16,000)	($39,000)
decreased notes payable	$15,000	$15,000	$15,000
increased working capital	$51,000	$67,000	$162,500

Figure 11.7 Sources and Uses of Cash Statement

Adapted from JIAN BizPlan*Builder*™, with permission.

Many people break out a simple cash flow statement by month or quarter for the first year or two (illustrated in Figure 11.8) which gives a clearer picture of your net cash balance over time (which is of significant interest if you want to quickly see that you can make every payroll). Data from June through November are suppressed to improve readability. Earlier business plans often include only a cash flow projection.

	Jan.	Feb.	Mar.	Apr.	May	Dec.
Beginning Cash Balance	$10,000	$44,174	$32,317	$19,459	$10,950	$46,061
Cash Receipts						
sales	$31,500	$34,650	$38,115	$41,927	$46,119	$89,873
interest income	$44	$193	$141	$85	$48	$202
total cash receipts	$31,544	$34,843	$38,256	$42,012	$46,167	$90,075
Cash Disbursements						
accounts payable	$15,800	$16,748	$17,753	$18,818	$19,947	$29,993
advertising	$1,200	$1,200	$1,200	$1,200	$1,200	$1,200
commissions (10% of sales)	$3,150	$3,465	$3,812	$4,193	$4,612	$8,987
salaries	$2,500	$2,500	$2,500	$2,500	$2,500	$2,500
other expenses	$3,100	$3,162	$3,225	$3,290	$3,356	$3,854
tax payments	$1,500	$1,500	$1,500	$1,500	$1,500	$1,500
total cash disbursements	$27,250	$28,575	$29,990	$31,501	$33,115	$48,034
Net Cash from Operations	$4,294	$6,268	$8,267	$10,511	$13,052	$42,040
equipment lease	$125	$125	$125	$125	$125	$125
equipment purchase	$1,995	$0	$3,000	$895	$18,500	$4,400
office lease	$3,000	$3,000	$3,000	$3,000	$3,000	$3,000
short-term loan repayment	$15,000	$15,000	$15,000	$15,000	$15,000	$15,000
sale of stock partnership units	$50,000	$0	$0	$0	$0	$0
proceeds of bank loan	$0	$0	$0	$0	$25,000	$0
Net Cash Balance	$44,174	$32,317	$19,459	$10,950	$12,377	$65,576

Figure 11.8 Simple Cash Flow Statement

Adapted from JIAN BizPlan*Builder*™, with permission.

While the cash flow statement shows actual cash requirements, the sources and uses of funds includes noncash expenses like depreciation and therefore shows more clearly cash availability and expenditure.

Again, it is helpful to show your reader this data in graphic form, as shown in Figure 11.9.

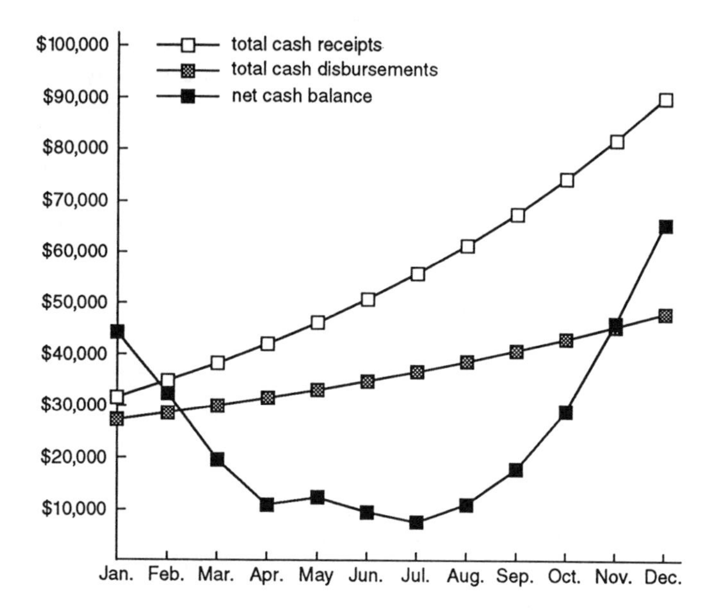

Figure 11.9 Cash Flow Projection Chart

JIAN states that:

> The cash flow projection incorporated in BizPlan*Builder*TM by most standards is a hybrid. Several categories are broken out to make it easier to project cash requirements based on decisions regarding advertising, sales commissions, salaries, equipment leases/purchases, and office lease. These expenses are the ones most often under direct management control. They are the ones (in addition to cost of sales) where funding will directly apply, and among other things, [which] you must convince your investors to fund.
>
> Outside of operating cash flow, your money and that of your investors is entered in beginning cash balance, sale of stock, and

proceeds of bank loan. Work with these numbers to maintain a positive net cash balance; a rule of thumb to use is to raise between 50—100% more cash than the cash flow statement indicates you will need.

BREAK-EVEN ANALYSIS

Since becoming profitable as soon as possible is the most important goal you can make if you want to optimize the valuation of your business, you and your investors will want to know when your business will break even. Break-even occurs when sales equal costs. A break-even analysis determines this point by computing the volume of sales at which fixed and variable costs will be covered. All sales above the break-even point produce profits; any drop in sales below that point will produce losses. A break-even analysis is a rough but important gauge of the volume of sales you need to attain profitability. Break-even for an imaginary company is shown in Table 11.2.

Table 11.2 Break-Even Analysis

	Per Month Optimistic	Per Month Realistic	Per Month Pessimistic
Fixed Costs			
administrative costs	$225	$563	$1,406
R&D investment	$2,800	$4,200	$6,300
selling costs	$2,100	$3,150	$4,725
total fixed costs (TFC)	$5,125	$7,913	$12,431
Variable Costs			
cost of goods sold	$4,490	$6,735	$10,103
total variable costs (TVC)	$4,490	$6,735	$10,103
Pricing & Unit Sales Variables			
selling price (SP)	$79.00	$79.00	$79.00
number of units (U)	600	500	400
variable costs per unit	$7.48	$13.47	$25.26
break-even unit volume	72	121	231
gross profit	$37,785.00	$24,852.00	$9,066.00

Adapted from JIAN BizPlan*Builder*™, with permission.

Fixed costs include such items as rent, management salaries, utilities, insurance, taxes, and depreciation. Variable costs include items such as supplies, outside labor, raw materials, production wages, advertising, and maintenance.

Again, it is important that you prepare a break-even chart similar to the one shown in Figure 11.10. Nothing is more frustrating than trying to see what numbers mean in tabular form. All modern spreadsheet programs support charting from your raw data. By example, each of the charts you see in this chapter was created in just a few minutes using the data in the example financial pro forma spreadsheets.

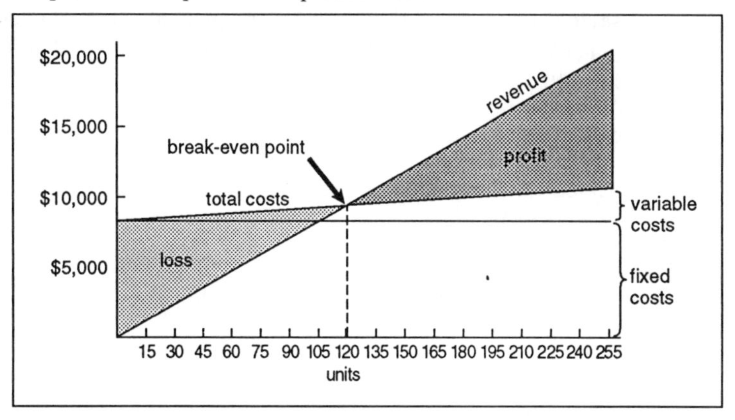

Figure 11.10 Break-Even Analysis Chart: Realistic Case

INVESTOR'S HURDLE RATE OF RETURN ON INVESTMENT

It is to your advantage to know what measures investors use in determining which companies they will invest in, how you can compute these measures, and how you can use them to evaluate and help sell your business plan. If your venture's rate of return on investment does not exceed your investor's hurdle rate (i.e., the minimum needed to be considered), your plan will not be given further consideration.

Basic Terms

Once you understand several related terms, you will have the vocabulary to play the investment game as it pertains to funding your start-up. While the subject of finance can get very involved, all you need to learn are the basic terms. They are all defined in relation to one another, and most of them boil down to one concept: discounting cash for the time-value of money. More advanced concepts such as the capital asset pricing model (CAPM; takes into account risk), the optimal capital structure (OCS; determines the optimum proportion of debt and equity), and the weighted average cost of capital (WACC; the average rate of return that capital investors expect the company to earn) can be left for graduate school.

Present value (PV) defines what one dollar is worth today (namely one dollar). More important, it is the value today of a future payment or stream of payments, discounted at some appropriate compound interest or discount rate. For example, the present value of $100 to be received 10 years from now is about $38.55, using a discount rate of 10% interest compounded annually. This is simply the time-value of money. If you were to put your savings in a riskless (before-tax) 10% annually compounded savings account, it should make no difference to you if you received $38.55 today or $100 in 10 years.

I = cash invested at the beginning of a holding period

PV = discounted value of a future payment or series of payments

Future value (FV) is simply the amount to which an investment will grow at a future time if it earns a specified interest that is compounded annually. It can easily be shown that the FV of an initial investment I compounded annually at rate i for n years is:

$$FV = I(1 + i)^n \quad [1]$$

= cash yielded at the end of an investment holding period

Cash-on-cash returns (COC) is the simplest measure of an investment return. It is widely used to describe venture capital investments.

$$COC = \frac{FV}{I} \quad [2]$$

As you can see, COC does not take into account the time-value of money. Therefore, investors often describe their performance goals in terms of COC and a period of time. For example, if you hear that an investor expects a 10 times return in five years, that means COC should equal 10 after a five-year holding period.

One can easily compute the actual annual compounded rate of return, i, achieved by a particular COC return over a period of n years from the preceding equations. Financial calculators or tables are usually consulted.

From Equation [1],

$$(1 + i)n = \frac{FV}{I}$$

Substituting Equation [2],

$$(1 + i)^{n} = \mathrm{COC}$$

$$(1 + i) = \mathrm{COC}\,\frac{i}{n}$$

$$i = \mathrm{COC}\,\frac{i}{n} - 1 \qquad [3]$$

Thus, in the example of COC = 10 in $n = 5$ years,

$$i = (10)\,\frac{1}{5} - 1$$

$$= (10)^{0.2} - 1$$

$$= 1.584893 - 1 = 0.584893$$

$$= 58.4893\%$$

Under the net present value method (NPV), the present value (PV) of all cash inflows from an investment is compared against the initial investment (*I*). The net present value (NPV) is simply the difference between PV and *I*.

NPV determines whether an investment is acceptable. To compute the PV of cash inflows, a discount rate called the *cost of capital* is used for discounting. Under this method, if NPV > 0 (or equivalently, PV > (*I*), the investment should be made.

Rate of return or rate (of return) on investment (ROI) is equivalent to the internal rate of return (IRR) needed for the net present value of one's investment to be zero. Any financial calculator will spit out that number. This is sometimes a difficult concept to understand. Working through a few examples in a finance textbook will make this relationship clear.

Thus, IRR is the discount rate at which the present value of the future cash flows of an investment equal the cost of the investment. When the net present values of cash outflows (the cost of the investment) and cash inflows (returns on the investment) equal zero, the rate of discount being used is the IRR.

When IRR is greater than the investor's required return on investment (called the *hurdle rate* in capital budgeting) the investment is acceptable.

An investment can profitably be made when IRR exceeds the cost of capital.

Investors May Examine Your Financials First

Some venture capital investors will assign a part-time junior assistant (usually an MBA student) to first examine your financial pro forma documents (informally referred to as your *financials*) before attempting to understand your business (i.e., your market, your product, or the strength of your team). This is done with the thought that if your numbers do not reflect the return needed to meet their ROI hurdle criterion, they would not invest in your start-up. For this reason alone, you must do your best in preparing your financial projections.

Because the investor's ROI depends on his or her percentage equity ownership and price per share paid (items yet to be negotiated), ROI does not fall out of any of the business plan financial pro forma projections.

You could compute ROI for a given investment scenario and include this in your plan, but most people do not. Including investor's ROI in your financial section forces you to make a presumption on the valuation of your company, and many investors prefer that you do not do that for them. While it might seem to make sense to include this information to prove that you know that this is a good investment, do not do so. If you want VCs' money, play by their rules. In oral presentations to VCs, be prepared to respond to questions concerning ROI.

Valuation

Investors want to calculate their ROI themselves based on their valuation calculations and using their methods, and they probably figure that in doing so they have a little edge on you. Calculating an investor's ROI is akin to computing someone else's income taxes without knowing their income (because investors will discount your projected revenues to a level they feel is attainable, and you will not know what discount they used). Because no two people would likely give your company the same valuation independently, calculating an ROI for your investors is indeed presumptive. Only when you understand how much your company is worth to your investors can you begin to estimate how much equity you should give up for your seed or start-up capital. Without this knowledge, you may just have to settle for a rule-of-thumb explanation for any proposed pricing, and your stock will likely be underpriced. Unfortunately, supposed rule-of-thumb valuations have funded more companies than you might imagine.

While you will not include ROI in your plan, you must calculate it to make sure your plan will not be rejected. It is surprising how many entrepreneurs neglect to compute ROI.

A company's ROI can be expressed as a percentage earned on the company's total capital (its common and preferred stock equity plus its long-term funded debt), calculated by dividing total capital into earnings before interest, taxes, and dividends. It should be easy enough to calculate ROI for a company. However, while this number might be an overall useful measure, for the individual investor it does not answer the question, "Should I invest in this company?"

When an individual investor computes his or her ROI, the thinking is, "How much will my return be in compounded annual terms, in so many years, when I cash out?" To figure this out, one needs to know:

- How much is paid per share at year 0?
- What can a share be sold for at year *N*?

From these two valuation-related numbers, you can see that ROI is the internal rate of return (IRR) needed for the net present value of one's investment to be zero. Any financial calculator will furnish that number.

- Given a target IRR, one can compute the purchase price per share needed for an assumed cash-out price and holding period.
- Given a target IRR, one can compute the cash-out price per share needed for a given purchase price and holding period.
- Given a target IRR, one can compute the maximum holding period one can tolerate in order to cash out at an assumed cash-out price for a given purchase price.
- IRR can be computed by setting any other target variable and making estimates of the others.

Your potential investors will attempt to compute your company's future valuation, figure in their required percentage ownership, and determine what would be a fair purchase price reflecting a judgment of current valuation.

Table 11.3 IRR for Given Cash Flow

year 0	year 1	year 2	year 3	year 4	year 5	IRR
($10)	$0	$0	$0	$0	$10	0.00%
($10)	$0	$0	$0	$0	$11	1.92%
($10)	$0	$0	$0	$0	$12	3.71%
($10)	$0	$0	$0	$0	$13	5.39%
($10)	$0	$0	$0	$0	**$14**	**6.96%**
($10)	$0	$0	$0	$0	$15	8.45%
($10)	$0	$0	$0	$0	$16	9.86%
($10)	$0	$0	$0	$0	$17	11.20%
($10)	$0	$0	$0	$0	**$18**	**12.47%**
($10)	$0	$0	$0	$0	$19	13.70%
($10)	$0	$0	$0	$0	**$20**	**14.87%**
($10)	$0	$0	$0	$0	$21	16.00%
($10)	$0	$0	$0	$0	$22	17.08%
($10)	$0	$0	$0	$0	$23	18.13%
($10)	$0	$0	$0	$0	$24	19.14%
($10)	$0	$0	$0	$0	$25	20.11%
($10)	$0	$0	$0	$0	$26	21.06%
($10)	$0	$0	$0	$0	$27	21.98%
($10)	$0	$0	$0	$0	$28	22.87%
($10)	$0	$0	$0	$0	$29	23.73%
($10)	$0	$0	$0	$0	$30	24.57%
($10)	$0	$0	$0	$0	**$31**	**25.39%**
($10)	$0	$0	$0	$0	$32	26.19%
($10)	$0	$0	$0	$0	$33	26.97%
($10)	$0	$0	$0	$0	$34	27.73%
($10)	$0	$0	$0	$0	$35	28.47%
($10)	$0	$0	$0	$0	$36	29.20%
($10)	$0	$0	$0	$0	$50	37.97%
($10)	$0	$0	$0	$0	**$70**	**47.58%**
($10)	$0	$0	$0	$0	**$100**	**58.49%**
($10)	$0	$0	$0	$0	**$150**	**71.88%**

Table 11-3 shows an investor's IRR for several investment scenarios. Your task is to show that your company will generate such good profits so rapidly that the investor's hurdle rate of return on investment criterion can be met without needing to acquire an unreasonable percentage of your company in the seed or start-up round.

Table 11-3 also shows cash flows from the investor's pocketbook. Assume that $10 is paid per share in year zero. For simplicity, no additional cash exchanges hands during years one through four. In year five, anywhere from $10 to $150 is returned to the investor at cash-out time. The investor's corresponding IRR is shown in the right-hand column.

How Big Does ROI/IRR Need to Be?

Obviously, investors would like to hit all home runs. Ideally, an investor would get 10 times his or her money back in five years, as shown in Table

11-3; a yield of 58.49% compounded return. In reality, though, the numbers are lower. While VC fund returns may be single digit, VCs still will not invest in any single venture unless they see five to ten times returns (40—60% annually) due to the fact that some investments will fail and others will not perform as well.

A number of years ago, venture capital funds were expected to get 25% returns, meaning that they would triple their investment in five years. Returns of 40—50% were not uncommon, as reported by the *Wall Street Journal*, June 20, 1991. More recently, single digit returns are more usual, returning anywhere from 1.4 to 1.8 times an initial investment over a five-year period. Investors are often finding that they need to stick with a start-up longer than expected. You can imagine that an additional couple of years' holding period drastically lowers returns.

Acquisitions are increasingly the exit vehicle most often available for venture capitalists, as opposed to initial public stock offerings. Unfortunately, cash-on-cash return (i.e., return not discounted for time value) for an IPO is typically seven times the original investment ($70 in the year 5 column in Table 11.3, yielding a 47.58% IRR), whereas an acquisition typically delivers two times the original investment ($20 in the year 5 column in Table 11.3, yielding a 14.87% IRR).

Actual returns earned by venture capital firms are less than might be imagined. Many of the funds formed in the early '80s returned 25% annually or more. The generation of funds formed from 1983 through 1987, known as the *lost generation* of funds, is averaging less than 10%. Many other venture funds formed after 1983 are permanently under water (i.e., are worth less than what they started with) and many will be lucky to earn a 5% return.

EXIT STRATEGY

More and more entrepreneurs are including a section in their business plan that recommends a preferred exit strategy: for example, acquisition by a large company that is also a strategic partner. This is a change from a few years earlier when few gave thought to whether (and if so, how) the business would be sold. The reason for this new awareness is probably that more start-up entrepreneurs see starting a business as a means to an end, not an end in its own right.

PRIVATE PLACEMENT MEMORANDUM

You may see a variation of the general business plan designed for subsequent rounds of funding for ongoing start-ups from the venture capital community and from sophisticated individual investors. In these cases, the plan is targeted toward these sophisticated investors and serves multiple functions:

- It embodies major sections of the company's general business plan.
- It is a selling document, and thus supposedly represents reality.
- It is a legal document, meaning that it is the basis for an investor's decision. Because charges of fraud and securities violations can arise from misleading statements or omissions in such a document, it will be a more honest accounting of the business.

Such a business plan will often be entitled a *private placement memorandum*. Here will be described the number and kinds of shares (equity) and notes (debt) authorized, issued, and offered, and at what prices. The legal structure of the business is also described in more detail. Copies of key manager employment contracts (if they exist) will be included, too, along with remuneration schedules of key managers. Terms of stock option arrangements for key managers will be precisely described. Actual past financial statements are presented. Finally, names of accountants, attorneys, and bank officers will be set forth.

In conclusion, as you can see, a well-planned business requires a lot of homework. If you understand all the issues explained in this chapter, you are indeed very well prepared to proceed. If your start-up does not have a well thought-out business plan, you may be taking a big risk. You will have presumably generated more information by this time than you thought possible and will know whether to proceed.

Business Plan Writing Aids and Financial Planning Aids

There now exist good computer planning aids that will minimize your writing and financial planning efforts. BizPlan*Builder*™ by JIAN, Tools for Sales, along with Tim Berry's *Business Plan Toolkit*™ by Palo Alto Software, are worthy of your consideration, and both are available for either the PC or the Macintosh. Each of these provides a complete business plan text template you can use with almost any word processor to guide you through the construction of your plan. While specific wording with blanks is often provided, you do not have to use the templates' words; you can instead just substitute your own. In other places you will find well-marked reminders to make sure you have considered important points. When you are satisfied that you have included the information needed, or determine that a particular section does not apply to your business situation, merely delete the remainder text.

For more information, contact the following:

- JIAN, Tools for Sales, 127 Second Street, Top Floor, Los Altos, CA 94022, (800) 442-7373, (415) 941-9191, $129.00.
- Palo Alto Software, 260 Sheridan Avenue, Suite 219, Palo Alto, CA 94036, (800) 229-7526, (415) 325-3190, $149.95.

BizPlan*Builder*™

This comes with a complete manual on how to write a business plan along with extensive suggestions for wording to be used in your plan. The following is a sample of this tool's excellent text-writing aid for the executive summary.

Executive Summary

- This section is an abstract of your company's present state and future direction.
- It is usually written after the plan is completed because it gives readers an overview of your business and it indicates how your business plan is organized.
- Edit to about two pages.

In 19__,_______(your group, company, product developers) were formed to _______(purpose of activities.

Now, (Company) is at a point where ________.

- Corporate mission statement covering the line of products and services—what kind of company do you want to be?

Background

For many years people have _______.

- How many people are managing to do without.
- How and where a similar product or service is now being used.

We have just started/completed the development/introductory phase of (product/service)—a novel and proprietary ____(e.g., soap for cleaning vinyl). Our operation was producing____(sales, units, products) by 19__, and has operated at _____(financial condition—profitable, break-even, etc.) ever since. Revenue projected for fiscal year 19__ without external funding is expected to be $_____. Annual growth is projected to be___% per year through 19__.

Concept

The state of the art/condition of the industry today is such that ______.

- Explain your place in the state of the art.
- Description of your product or service.
- Desirability of your product or service.

Compared to competitive products (or the closest product available today) our (product/service) ______.

The ability to _____ is a capability unique to (Company)'s product/services.

- How would your customers compare your product with those of competitors?
- Advantages product or service has—its improvements over existing products or services.

Our strategy for meeting the competition is ______ (lower price, bigger/better—your unique selling proposition).

(Company)'s target market includes _____ (types of customers). (Company) is rapidly moving into its _____ (marketing phase).

This approach is generating a tremendous amount of interest throughout our industry.

_____ follow-on products/services, _____ (product) is a _____ and is especially useful to _____ (prospective customers) who can now easily _____.

Other products/services include _____.

All products from (Company) are protected by the trademark and copyright laws, and _____ (patents, etc.).

Responses from customers indicate that _____(product/service) is enjoying and excellent reputation. Inquiries from prospective customers suggest that there is considerable demand for _____. Relationships with leading OEMs (original equipment manufacturers), retailers, major accounts, manufacturers, and distributors substantiate the fitness of (Company) for considerable growth and accomplishment.

Objectives

- Near term and long term

Our objective, at this time, is to propel the company into a prominent market position. We feel that within ___ years (Company) will be in a suitable condition for an initial public offering or profitable acquisition. To accomplish this goal we have developed a comprehensive plan to intensify and accelerate our marketing activities, product development, services expansion, engineering, distribution and customer service. To implement our plans we require an investment of $____ for the following purposes (choose the activities pertinent to your goals):

- Build manufacturing facilities and ramp up production to meet customer demands.
- Maximize sales with an extensive campaign to promote our products/services.
- Reinforce customer support services to handle the increased demands created by the influx of new orders and deepened penetration into existing accounts.
- Augment company staff to support and sustain prolonged growth under the new marketing plan.
- Increase research and development to create additional follow-on product/services as well as to further fine-tune our competitive advantages.

Management

Our management team consists of ___ (how many) men and women whose backgrounds consist of _____ years of marketing with _____ (*Fortune* 500 company names always look great here), years of corporate development with _____ (more *Fortune* 500 company names look great here too), ______ people with ____ years of engineering and design with ______.

- Actually, any good company backgrounds pertinent to your management team's functions are good references to demonstrate a solid background and assure a higher probability of future success.

Marketing

_____ (research firm, industry report, trade journal study) research projects a worldwide/nationwide market for _____(product/service) to be approximately $____ by the end of 19__. Conservative estimates suggest (Company)'s market share, with our intensified and accelerated marketing plan, product/service development, manufacturing, and customer service would be about ___&—generating $____ by the end of 19__.

- Describe the projections and trends for the industry or business field.

The fundamental thrust of our marketing strategy consists of ____ (unique selling basis).

We intend to reach _____ (market segment) by _____ (marketing/sales/promo tactics).

- Who are your customers? Where are they and how do you reach them?
- Are they buying your product/service from someone else?
- How will you educate customers to buy from you?

Our company can be characterized as a _____ (the business and image for customers to see).

(Company) enjoys an established track record of excellent support for our customers. Their expressions of satisfaction and encouragement are numerous, and we intend to continue our advances in the ______ (marketplace) with more unique and instrumental _____ (product/services).

A partial list of customers includes:

____

____

Also, ______ prospective clients presently evaluating _____ (product/services) for use are _____ (actual customers).

- List customers in customers section

Finance

- Briefly forecast financial expectations.

In ____ years we will have _____ (achieved goal) and our investors will be able to ____ (collect their return on investment).

Adapted from JIAN BizPlan*Builder*™, with permission

Business Plan Toolkit

Business Plan Toolkit for the Macintosh contains a much less lengthy suggestion of text segments for you to go through, but does provide a useful HyperCard utility for collecting and sorting your thoughts on electronic notecards. When you are finished making your electronic cards, they can be automatically collated into one text file for polishing and merging with other financial and graphic data. They HyperCard utility might be attractive to regular HyperCard users, but the small editing window and scantiness of text suggestions may be somewhat of an annoyance, especially on a large screen Macintosh. A screen shot of this tool's HyperCard aid for writing the executive summary is shown.

You can simply click with your mouse on the folder tab buttons to open any of the business plan sections (Summary, Company, Product, Market, Strategy, Organization, or Financial) and enter a new HyperCard prompting script such as that shown for the executive summary.

Business Plan Toolkit's sample text files can also be used to guide your writing process without the HyperCard aid, as shown (again, this is an example for the executive summary).

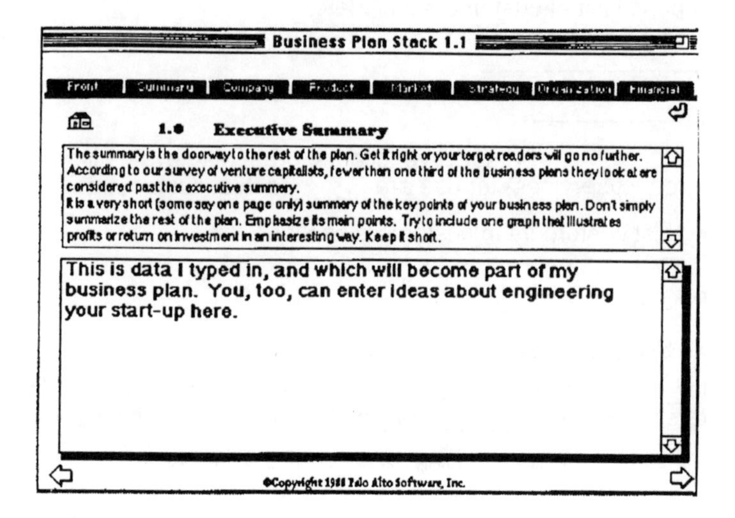

Figure 11-11 Business Plan Toolkit Screen

Adapted from Business Plan ToolkitTM, by permission
from Tim Berry, Palo Alto Software.

Executive Summary

- Acme Widget Company (AWC) produces industry-specific widgets from standard widgets. It has its main offices in Upscale, CA, and maintains a small manufacturing and assembly plant in Standard, CA. Its products are sold through distributors and direct-response marketing.
- This business plan is part of our regular business planning process. We revise this plan every quarter.
- In the next full year we intend to develop two new widget products and to improve revenues to more than $250,000 monthly.
- Our keys to success and critical factors for the next year are, in order of importance
  Product development
  Sales to dealers in volume
  Financial control and cash flow planning
- AWC is a growing concern that has been profitable for the last five years. Figure 1 illustrate highlights of our financial performance over the last tto more than $440,000, while net profits are us to more than $50,000.

Also included in *Business Plan Toolkit* and BizPlan*Builder*™ are spreadsheet templates for your financial pro formas. While you can certainly make your own, why not take advantage of good tools created by others? "The right tool for the job" is an expression that applies her. Do not get stingy with a hundred dollars now when time and completeness are so critical for your success. You are likely to find these packages for sales at a discount software stores for as low as $79.

Ronstadt's Financials also markets financial planning tools that are designed more for managers and business professionals who cannot afford to spend time learning to use or write formulas and format their own spreadsheets. Ronstadt's Financials is available from Lord Publishing, 49 Eliot Street, Natick, MA 01760 or PO. Box 806, Dover MA 02030, (508) 651-9955. It costs about $119 and includes Ronstadt's book, *Entrepreneurial Finance.* When ordering, specify 3.5" or 5.25" diskette. This package is designed for the PC only. You use it by entering your business' operating assumptions on a single form. The program automatically generates cash flow statements, balance sheets, and income statements.

In order for you to get a better feeling for the value of the financial planning parts of the two packages reviewed, a few examples have been included. Both are very well done.

BizPlan*Builder*™'s, templates consist of both a large integrated template that includes all the schedules in one and individual nonlinked schedules. These represent two fairly different ways to construct your financial pro formas. In the integrated template method, one changed entry automatically updates all the other fields (wherever it is used), but only within this one very large integrated template. That is, rather than using external links between different smaller spreadsheets, there is just one large integrated spreadsheet. To use this spreadsheet you need a fully functional product like Lotus 1-2-3 or a compatible spreadsheet such as Excel. The integrated spreadsheet is recommended unless you spreadsheet program will not support it, in which case you can use the individual nonlinked spreadsheets. However, since you will be using the tool almost daily in your business and there is absolutely no time for errors or delays when you are writing your business plan, it is strongly recommended that you obtain a modern spreadsheet product.

Finally, BizPlan*Builder*™'s templates are built on an assumption-driven model with slots to fill in for your expected starting values and growth rates for all key variables, which can be an exceptionally useful feature.

Business Plan Toolkit's templates consist of an iterated template externally linked to many smaller input schedules; thus, all estimates and changes made in any individual worksheet schedule (sales, for example) show up in the integrated document. In addition, macros are provided to automatically open spreadsheets and charts. For some, the *Business Plant Toolkit*'s multiple-linked-document product represents a somewhat better, more modular approach than BizPlan*Builder*™'s single large integrated template. *Business Plan Toolkit* does require Lotus 1-2-3, Microsoft Excel, or a compatible spreadsheet on the PC, or Microsoft Excel or Works, Lotus 1-2-3, or a SYLK compatible spreadsheet on the Macintosh. At the time of writing, *Business Plan Toolkit*'s templates were based on an older version of Excel (revision 1.5, which still works with newer versions), and a new product release was in the works.

In conclusion, if you are writing your first business plan, then owning one of these two tools is essential.

More experienced entrepreneurs might already have their text and financial templates from previous start-ups, but they would still benefit from the ability to do an on-line check or completeness with either one of these excellent tools.

A third useful package is one that has been adapted from Michael O'Donnell's *Writing Business Plans That Get Results*, a highly recommended workbook. His book has also been marketed as a business plan writing aid by Ronstadt's Financials under the title *Business Plan for Start-ups*. Their version provides word processing software for O'Donnell's book and includes many commonly asked questions.

To obtain this business plan writing diskette, sent $14.95 to Lord Publishing, 49 Eliot Street, Natick, MA 01760 or P.O. box 806, Dover, MA 02030, (508) 651-9955. Specify 3.5" or 5.25" floppy.

One Venture Capitalists Perspective on Plan Emphasis

Addressing the Silicon Valley Engineering Management Society in 1990, Michael Moritz, partner of Sequoia Capital, a Menlo Park, CA firms, stated some interesting opinions:

> In selecting an investment it's all in the timing. Investments in word processing software, for example, a few years before its time, resulted in big loses, whereas at a later point in time they were attractive.
>
> We ask where is the market, where are the customers [with discretionary funds], where are the [customers'] wallets, what is the product [and how long will it take to ship, especially for software?], and who is the management?

It is interesting that this statement puts less emphasis on management than most. Management and markets, not necessarily in that order, are usually listed as the two most important factors in evaluating a business opportunity.

> First, the market must be right! We want to achieve scale in a few years. Management must be of high quality, and we must see an exit strategy.
>
> We look for unexploited niches [networking utility software for business was mentioned as on good example].
>
> We want to see a substantial market potential, a proprietary product or service, existing channels of distribution, and high gross margins.

Moritz cited software as having good gross margin potential; hardware has less, since disk drives look unattractive at 28% gross margin (unless one captures over 50% of the market share). Radius at the time was an exception, having 37% gross margins.

> After the last financing round, 25-30% of the stock should stay with management.
>
> Our objective is to turn $1 million into $10 million.

We will typically spend one to two months on due diligence before we invest in a start-up.

Different areas are hot at different times; we like communications, semiconductors, software [prefer existing software], computers and associated peripherals, medical- and health-related technology [instrument] companies, and biotechnology.

Chapter 12
FUNDING ISSUES

"Business? It's quite simple.
It's other people's money."
—Alexandre Dumas

INVESTMENT CRITERIA

Investors determine how much money they will invest based on your minimum needs, considerations of return on investment, and considerations of risk. Obviously the best opportunity with the lowest risk will attract the most funds. This is why your business plan must reflect your very best effort, and why you should have already made progress in building your management team. You are selling an opportunity, and the evidence of this opportunity and your potential for success is reflected in the essential elements of your business plan. A marketable business plan will clearly reflect:

- a unique market/business opportunity
- a complete and experienced management team with a solid track record
- attractive markets and a high likelihood of selling and distributing successfully to identified customers
- sound plans and the technological basis for developing and manufacturing the proposed products
- a clear vision and an operations plan for carrying out the business
- clear financial business objectives and an understanding of the funding requirements to make the venture successful (largely the subject of this chapter)

CHEAP START-UPS ARE FINISHED

Unfortunately, the days are past when you could start a high-technology company in your garage or spare bedroom and grow to be a leader in your industry. Stories beginning, "Founded in 1939 in a Palo Alto garage by Stanford electrical engineers William Hewlett and David Packard..." are history. Increased competition demands a more professional approach if you are to successfully compete for declining venture capital and other sophisticated investment resources.

A well-known general partner of a prominent Silicon Valley venture capital firm recently stated, "There will be no more successful stand-alone start-ups." By this, he meant that because competition is now so strong and so global, a successful start-up must from its beginning have plans for strategic partner relationships with customers (for product specification and distribution), manufacturers (for low-cost manufacturing), and the government (for plant construction and so forth).

He concludes,

> The rules have changed. While the pace of start-ups has actually turned up, we see more experienced teams starting ventures. Start-ups need at least two initial venture capital investors plus plans for follow-on financing. The big opportunities—and there are big opportunities, such as in pen computing, the life sciences—are still there.

LOOKING FOR SEED CASH

Your business plan pro forma financial documents should tell you how much cash you will need to get your venture to the break-even point, i.e., where your revenues will balance your expenses. However, this amount of cash may not represent what you should have to start. Realistically, you may have trouble raising that much seed cash. And, if you did raise that much cash, you might have to give up too much equity to do so. Even if you raised enough cash to take you to break-even, you would need even more cash to finance growth. So how much money should you seek to seed your venture, and from whom?

HOW MUCH MONEY?

It really is impossible to specify an exact number here without knowing more about your business plans. Many small software companies are entirely funded by custom development contract work from one or more of their customers. This strategy gets a prototype built without giving up any equity. Usually, though, additional funding will be required. It is extremely important that you never run out of cash; that would be your worst nightmare.

Ideally, you would like to have enough cash to operate for at least six months (see your cash flow statement). You do not want to interrupt your development schedule any earlier than necessary to go fund raising. Realize, however, that you will always be raising funds until your company is self-sustaining. You can make a good rough estimate of how much you will spend monthly based on head count alone. Unless you have better regional and industry statistics available, use the rule of thumb that a business will need to spend about $10,000 per month per employee including all overhead costs. For high-tech companies in high-cost geographic locations with large capital equipment needs, $20,000 would be a more typical target. Since you will likely have no significant income during your seed and start-up phases, your cash outlays dictate your cash needs. Therefore, a six-person start-up operating for six months, at 60% salaries, would consume about

$$6 \times 6 \times \$10{,}000 \times 0.6 = \$216{,}000$$

This estimate is consistent with the fact that a large fraction of private investor initial deals are for $50,000 to $300,000, and that most seed venture capital funds typically invest from $200,000 to $500,000 at the initial stage of a new company. Several additional rules of thumb can be found in literature, which may be useful:

- Have enough cash to attract key employees to the business and to look good enough to prospective creditors and landlords that you can rent space and equipment. Be able to make promises you can keep concerning payroll, taxes, and rent for the near future.
- If funded entirely through personal funds and funds from relatives, be able to show a bank balance of at least $100,000 at the formal launch of the venture.
- If funded by an angel or other sophisticated investor, try to raise a minimum of $300,000 before the launch.
- If funded by venture capitalists, try to show a bank balance of $500,000 to $1 million at the formal launch of the venture.

SEED, START-UP, AND SUBSEQUENT FUNDING ROUNDS

Your seed round financing should take you through the point where you can prove your product concept. These funds may involve product development, but they rarely involve initial marketing.

Although your start-up round financing will probably be a separate round following seed, ideally the two are combined so that you can concentrate on executing your plan instead of constantly looking for funding. Because the recent trend is to parcel out even smaller chunks of cash, you would be very lucky to secure a combined seed and start-up round these days. Start-up round funds are used for product development and initial marketing, assembling the key management team, perhaps preparing more detailed business plans, completing market studies, and generally preparing to do business.

First-stage early-development funds are then solicited to initiate commercial manufacturing and sales. You probably will not be profitable in this stage.

Expansion financing for your second-stage, third-stage, and fourth-stage financing steps follow the start-up phase.

THE NIGHTMARE: RUNNING OUT OF MONEY

While you hope to get each round funded in a timely fashion and for an adequate amount, there will be times when you are close to running out of cash and unable to make a payroll. One reason you want to structure your business for rapid profitability is to avoid this nightmare. If you run out of cash, you will be faced with the dilemma of trying to raise funds under duress while trying to keep the doors open. To avoid this, you need to raise enough funds, get profitable fast, and keep looking for money.

WHERE TO GET MONEY

Savings, Mortgages, Friends, and Relatives

A study of 600 high-tech firms by Edward Roberts of MIT shows clearly that personal savings provide the primary source of seed capital (see Figure 12.1). John Ward at Loyola University's Graduate School of Business states that parents constitute the largest single source of start-up capital in the country. Most likely, you will be risking your savings, and those of your loved ones, to launch your new business. Sandra Kurtzig, the founder of ASK Computer Systems, financed her $400-million business with a $2,000 commission check from a previous employer and a loan from her father for $25,000.

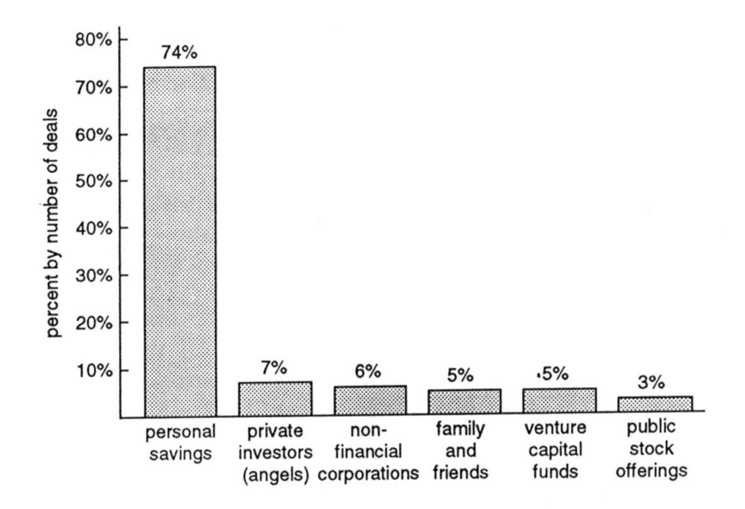

Figure 12.1 Primary Sources for Seed Capital for High-Tech Companies

The most successful entrepreneurs (those whose companies grow most quickly), however, initially had money from venture capitalists and angels. Once again it is emphasized that you try to start your business with a team. Find members who can help you attract investment funds.

Accountants and family business consultants will tell you that if you decide to borrow money from relatives, the only hope of making the loan work lies in being very explicit about the conditions of the deal in writing. Popular literature (*Inc.* magazine, *Success*, etc.) is littered with tragic stories of families torn apart from misunderstandings about money loaned to business start-ups.

Customers and Suppliers

A powerful source of capital in your seed and start-up stages can be your first customers. Getting customers to pay fat deposits up front with their orders (in return for discounted prices, for example) can relieve you of the need to raise substantial funds. Alternatively, instead of a cash deposit, you might request that your customer give you a letter of credit for a down payment which you could take to a bank and borrow against, thereby enhancing your funds and credit history simultaneously. Early customers also can provide an important revenue stream to your business if you provide them with services associated with the use of your product. Income from such services can be used to finance the continuing development of your products. At some point you might want to sell off such income substitution branches of your business to facilitate more product-oriented growth, but meanwhile, services can provide essential sources of funds. ASK Computer Systems used service bureau income to sustain itself in its early days. Largely as a result of this service income, ASK never needed venture capital.

Negotiating extended payment terms to your suppliers until your customers pay you can provide an additional cash cushion.

Angels

Angels are wealthy, private-individual investors who work with start-up companies, often at the seed-level stage.

Guy Kawasaki, in *Selling the Dream*}, describes angels as "people who share your vision and provide *wings*, such as emotional support, expert advice, and sometimes money—as a mother bird uses her wing to shelter her young."

Angels may be doctors, lawyers, other professionals, or successful entrepreneurs and businesspeople. They typically seed start-ups with a few tens of thousands of dollars up to hundreds of thousands of dollars. Angels often are in a position to give you good business advice, usually as members of your board of directors. Also, they will often inject funds into your business on more favorable terms than some venture capital firms. You will find angels in all walks of life, and it is often up to you to structure and propose a deal to them. Most angels, unfortunately, will not be in the best position to introduce you to the next round of investors.

By accepting angel money you will have started your business with a knowledgeable resource and minimal dilution. However, this often leads to the venture capital catch-22 problem (discussed after the venture capital section of this chapter). Some angels repel venture capitalists; the venture capitalists often prefer not to invest along with certain people, so be careful that your early angel is not an albatross.

Despite all the attention paid to venture capital firms, angels back more than 30,000 new and emerging businesses a year with about $10 billion, while venture funds back only about 2,000 with $2 billion, says William E. Wetzel, professor of management and director of the Center for Venture Research at the University of New Hampshire. Figure 12-2 illustrates how important angel capital is.

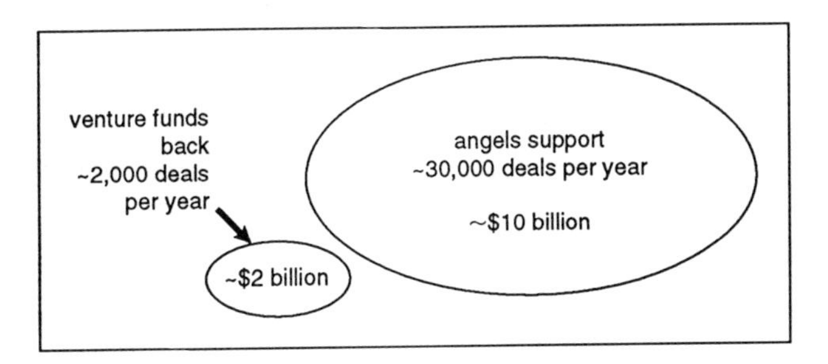

Figure 12.2 Angels versus Venture Capital Support

However, angels, while they may be easier to catch, can also be harder to please. Investors must typically be prepared to inject additional funds equal to double or triple their initial investments, and they will have to ride the company through rough times. Many angels do not have the investment experience to realize that these things will happen, while venture capitalists are almost always more experienced, their funds more heavily financed, and more able to adapt. Unlike venture capitalists, who may have worked with start-ups for years, inexperienced angels often cannot provide essential business advice. Venture capitalists are in tune with the specific markets they invest in. However, seasoned angels know what to expect also.

Angels who overprice seed deals may seem a delight to you, but to a venture capitalist looking at your start-up round, that may be reason enough to pass you up. If a venture capitalist does come in for your second round of financing and then reprices the deal, your angel will get diluted more than expected. If you anticipate this, but a nonsophisticated angel does not see it, fix that angel up with more shares if necessary.

Private Stock Offerings

As you investigate approaching a few angels for seed or start-up capital, you may be tempted to try to sell stock in your company to an even larger number of individuals. A private stock offering is sometimes made through the drafting of a Rule 504 private-placement memorandum that allows you to raise up to $1 million without too much trouble from the Securities and Exchange Commission. If you attempt this, you most definitely need to work with an experienced lawyer. Table 12-1 summarizes the major exemptions to registration with the SEC, along with comments on how they work.

Table 12.1 Summary of Exemptions to Registration with the SEC

	Exemption Limit	Limit Number of Investors	Documentation Required
SEC Rule 504	$1 million	none	Disclosure document must be cleared by one or more U-7 states; SEC Form D must be filed aftter sale.
SEC Rule 505	$5 million	none if all investors are accredited[a]; 35 if nonaccredited	SEC Form D if investors are accredited; if not, form S-18
SEC Rule 506	none	35 experienced[b] investors; no limit on accredited investors	For S-1 for experienced investors; Form D for accredited investors
1933 Act 4(2)	none	fewer than 25	whatever attorney deems necessary to protect exemption
1933 Act 3(a)(11)	none	none if all reside in the same state	Varies from state to state. Company must keep good records on investors and use of proceeds to protect exemption.

[a] *Accedited* investors are institutions or individuals with at least $200,000 in annual adjusted gross income or with a net worth of at least $1 million. The $1 million net worth does not include the worth of a personal residence.

[b] *Experienced* investors are people capable of evaluating the merits and risks of a prospective investment.

Source: Drew Field/Direct Share Marketing, San Francisco. Reported in *Inc.*, December 1991.

Fortunately, in early 1992, SEC chairman Richard Breeden proposed relaxing the rules to help small companies raise capital. Under these new proposals:

- Small companies seeking seed capital could issue up to $1 million in unrestricted securities per year.

- Companies could canvass the public for interest in the securities before making the offer.
- Regulation A, which allows companies to use a simplified disclosure document for limited public offerings that are directed mainly toward big investors, would be raised from $1.5 to $5 million. Also, information could be published to gauge investor interest before filing the offering.
- New, simpler registration forms would be created based on the size of the company rather than the size of the offering.

About 21 states have adopted a little-known funding process for small businesses known as the SCOR (Small Company Offering Registration) process. Also known as ULOR (Uniform Limited Offering Registration), SCOR enables small companies to go public and raise up to $1 million with less difficulty and expense than a traditional securities offering usually involves. The *Wall Street Journal*, January 21, 1992, reported that only a few dozen companies have tried to use SCOR. Washington state was most active. While SCOR offerings suffer from the absence of an active aftermarket for the underlying securities, the process is reported to work well where there is already a large group of customers or employees who are potential investors. According to the North American Securities Administrators Association, the following states have adopted the SCOR process.

Alaska	Massachusetts	South Dakota
Arizona	Mississippi	Tennessee
Idaho	Missouri	Texas
Indiana	Montana	Vermont
Iowa	Nevada	Washington
Kansas	North Carolina	Wisconsin
Maine	North Dakota	Wyoming

Selling stock in your start-up company is a very difficult thing to do; also, your stock will have no market, and you and your new company will have little reputation to attract such investors. Although many firms exist for the sole purpose of helping you make these private placements, even they tend to have limited success for the average engineer starting his or her first company. Save such excursions for your second start-up when you become famous from your first success, or at least wait a year or two until your company has a track record. At that point, a private placement service certainly can work. Private placements are a little like going public on a small scale. While this is an exciting idea, it is not practical for you now. Save your time and more than a little money.

Finally, do not make the mistake of thinking you can bypass or disregard the securities laws; the penalties are severe. Also, watch out for things like rules against advertising investment opportunities.

Venture Capitalists (VCs)

Venture capital firms are a kind of funnel that gathers money from limited partners and then distributes funds to a large number of carefully selected growing businesses. The venture firm managers, known as general partners, raise this money from pension funds, insurance companies, university endowments, corporations, wealthy individual investors, etc.

A typical VC fund raises $50 million from limited partners and invests in up to 35 companies for 10 to 12 years. (These days, 35 companies would be a lot to invest in, since VCs are having to stick with their portfolio companies much longer before they can cash out their investments.) By the end of that time, the companies will have either succeeded (returning a profit to the fund through a public stock offering or sale to a larger company) or they will have failed and been written off.

VC firms can make money in three ways: management fees. carried interest, and stock price appreciation after a company goes public. A standard 3% management fee once provided venture capitalists with a monthly draw. A 20% profit participation in each deal (called *carried interest*) was the real payoff. Appreciation from the public stock from companies going public before funds were distributed to the limited partners provided additional profits. More recently, however, the pension fund limited partners have negotiated management fees to just 1% or 2%, and while keeping carried interest at 20%, many now insist on a profit distribution immediately after each initial public offering.

Kurtzig, who grew her business completely without venture capital funds, has an opinion on venture capitalists:

> Nowadays it is venture capital that keeps you in business in the early going. But venture capital is impatient money, and I doubt many venture capitalists would have stuck with ASK as it continued to redefine itself in its first four years. Not having venture capital means never having to say you're sorry.

Few software companies really achieve the $50 to $100 million sales levels that VCs want to see in companies they will fund.

While returns of venture capital firms were once as high as 40—50% annually, single digits are more common now (reported by the *Wall Street*

Journal, June 20, 1991). In the five years that ended in 1990, venture capital funds collectively posted losses of an average of 3.8% a year (estimate by Morgan Stanley & Co., the *Wall Street Journal*, February 11, 1992).

A number of publications list current venture capital investments. The *San Jose Mercury News*, for example, runs excellent quarterly reports on the venture capital money tree. These reports show what is hot and what is not in the investment community, which can be valuable to you in your search for funds. For example, from the decrease in the number of new listings and the reporting of additional rounds on older deals, it was obvious in 1991 that investments in new computer-related seed deals were declining. To those keeping tabs, the trend was clear, as illustrated in Figure 12-3.

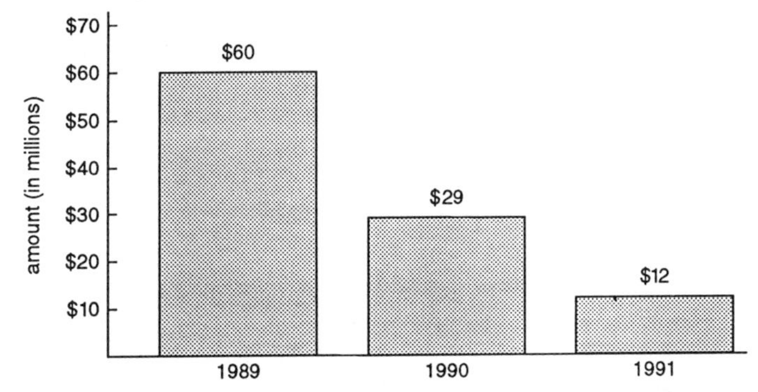

Figure 12.3 Computer-Related Seed Deals Closed

In 1991, only two dozen deals were consummated, each drawing an average of $500,000.

Nationwide, investments by venture capital firms peaked at $3.9 billion in 1987. That figure tumbled to about $2 billion in 1990, according to the market research firm Venture Economics, Inc., of Needham, MA. Reports in early 1992 showed a continued severe deterioration of venture capital disbursements to a level of just $801.4 million for 1991 (reported in *Venture Capital Journal*).

In 1991, the venture firms themselves attracted only about $1.34 billion from new investors (source: Venture Economics, February 1992). This figure is down from $1.8 billion in 1990 (source: the *San Jose Mercury News*, December 2, 1991 and February 24, 1992).

Only 35 partnerships raised money in 1991, a sharp decrease from 105 in record 1987 (source: Venture Economics, February 1992). Six big funds, including Oak Investment Management Co., Institutional Venture Partners, and Summit Ventures, accounted for more than half the total funds raised. In a dramatic unexpected turnaround, at the end of 1991, venture funds were expected to bring in as much as $2 billion in the first six months of 1992 (source: *Venture Capital Journal*). For perspective, a record $4.2 billion was raised in 1987.

Another positive trend for start-up engineers: early stage funds are back in vogue. Venture Economics says that of the total raised in 1991, 35% (nearly $470 million) was committed to partnerships investing in start-ups. That is a big increase from the $180 million or so committed to partnerships in 1990!

The most significant trend in Silicon Valley in the fourth quarter of 1991 was the amount invested in software companies—$62.4 million, more than any quarter since 1986 (source: the *San Jose Mercury News*, February 24, 1992).

The rate of money flowing into and out of VC funds will vary from year to year and the overall flow will be cyclical, as shown in the Figure 12.4 (source: Venture Economics, Inc., Venture Economics Publishing Co.).

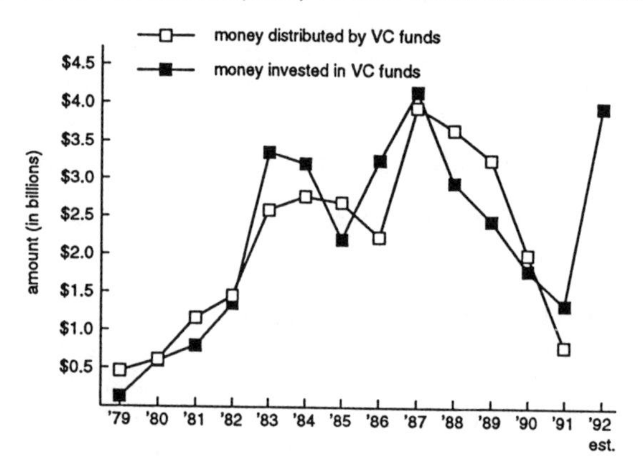

Figure 12.4 Venture Capital Funding Cycle

Other statistics from Ernst & Young show more important trends:

- Funding for computer hardware and semiconductors continues to shrink as these industries mature.
- Communications is an important sector with sharply increased demand for computer network systems and fiber optics.

- Venture funding for biotechnology and medical products declined as these companies obtained more money from initial public stock offerings.
- Software continues to attract the biggest share of venture funding, reflecting a strong demand for new programs and the small amount of investment needed to start a software business. (Obviously, software companies continue to provide the best paths for engineers wanting to start their own companies.)

As reported in *Venture Capital Journal*, July 1991, the sharp, pervasive reductions in venture capital disbursements are illustrated in Table 12-2.

Table 12.2 Recent Reductions in Venture Capital Disbursements

Industry	Disbursements in Millions of Dollars		
	Dollars Invested (1990 Q1)	Dollars Invested (1991 Q1)	Percent Change
commercial communications	23	0.9	-96
biotechnology	48	6	-88
medical/health care	94	19	-80
consumer-related	84	17	-80
software and services	108	62	-43

Seed disbursements fell more rapidly than the market as a whole. Only seven companies were the beneficiaries of seed investments in the first quarter of 1991 ($4.2 million), versus 21 in the first quarter of 1990 ($16.9 million). Likewise, start-up investments totaled just $9 million in the first quarter of 1991 versus $41 million in the first quarter of 1990. The huge declines ran through all industry categories. Software and services received the lion's share of the dollars disbursed.

Receiving any venture capital investment in early 1991 was quite an accomplishment. Ten companies pulled in a total of only $87 million; this represented almost one-half the entire amount disbursed. Three of the ten largest deals involved software companies.

Remember, the venture capital business is very cyclical, and by the time you read this book, venture funds could again be more widely available. It is essential that you keep abreast of these trends so that you do not waste time chasing improbable funding sources. This information will also be extremely valuable to you in setting realistic expectations for your start-up. There are many industry-specific publications that you might want to consult. *Computer Letter*, based in New York, covers the computer-related industries, for example.

You will find that venture capital investments are usually listed by the following categories:

- computers
- peripherals
- semiconductors
- communications
- biotechnology and medical products
- software
- miscellaneous

Very often a company founder finds it exceedingly difficult to raise funds, especially from sophisticated venture capital sources. Despite the connotation of the name, venture capital is not the reckless business of throwing money at speculative deals. A great deal of due diligence goes into evaluating an investment option, and risk is always carefully calculated and factored into the structure of investments. Investors want assurances that they will get a return on their investment. As stated previously, they want to see a management team that has performed well in the past and that will do so again in the future. They want to see you working in markets that are growing and in which you have distinct competitive advantages. They are interested in proprietary technology that offers that competitive advantage, but not technology for technology's sake.

Sophisticated investors also want to see product prototypes—evidence that you can produce what you say you can. In short, they want to bet on a horse that looks like it will cross the winning line. But how can you demonstrate this confidence without expending initial funds? Since you cannot, you get some angel's or relative's money and do your best, thinking that soon you will really impress the pros. Ideally, your initial seed funds will produce a believable product prototype and you will be ready to shop your business plan.

The catch-22 comes in when a sophisticated investor, now considering your plan, exclaims that "this looks pretty neat, it is too bad that you went ahead and got these unsophisticated (angel, relative) investors involved; that muddies the waters for us—they are hard to work with." It is a catch-22 since you often cannot get the attention of venture capital (new money) until you have produced something, and you cannot produce something without getting seed funds (old money), but venture capitalists often do not like to mingle with amateur investors.

There are ways out of this catch-22, however:

- The new money can cash out the early stage seed investors, but you should prepare your seed investors for that possible preference before they invest in your business. This is the cleanest move and makes sense. Your seed investors make a quick return and everyone should be happy.
- The new money can help convert the seed investors' equity position into a debt position, but this should be done only if that would match the seed investor's investment objective.
- The new money can convert the old money's preferred equity position into a common equity position. If this happens, make sure that your loyal seed investor is compensated for the new, riskier common position through the conversion of the preferred equity into a significantly larger (typically three to ten times) number of common shares.
- The new money can live with the old money—the venture capitalists and seed investors can share board seats and vote together. This last option is usually satisfactory when the seed investor is a veteran angel, but it is rarely satisfactory when inexperienced relatives and friends are involved.

It is best to get a seed-level venture capital firm involved in your business at the earliest date possible. It will have the networking connections to assist you in getting your second round of financing, and will see that all subsequent financing rounds are equitable. Getting a quality investment firm involved from day one, however, probably will require that you have formed a complete and experienced management team containing at least one individual who was successful in an earlier start-up. Again, if you try to do everything yourself, the growth of your business will likely be limited.

SHOPPING VERSUS SELLING YOUR BUSINESS PLAN

Shopping a plan in the venture capital trade means sending it out to too many investors at the same time. Because of the close-knit nature of the venture capital community, it is not at all unusual for these people to share information about potential deals. If you give your plan out to multiple venture sources at the same time and they know about each other, no one may give you the attention you are looking for since each wants to invest time on deals that they likely can close on if they become interested. Similarly, if your plan has been in circulation for a long time, it will be labeled shopworn. No one wants to spend time evaluating a plan that 100 other investors have passed on. This section is about how to sell your business plan without overshopping it.

Norman A. Fogelsong and Kenneth J. Kelley of Institutional Venture Partners offer the following formulas for how to get started in business and how to approach a venture capitalist.

Getting Started

- Plan on establishing a serious partnership with your investors and your management team that will last at least five to seven years.
- Make the firm decision to start a business; your start cannot be contingent on "ifs" and "maybes."
- Get the support of your family (this is absolutely essential).
- Establish very clearly your raison detre} (your unique contribution and business mission).
- Identify the market opportunity, develop the product definition, and build the founding team. (It cannot be emphasized enough how important building a team is.)
- Write the initial business plan.

How to Approach a Venture Capitalist

- You should first select no more than three to six venture capital firms whose investment profiles fit your start-up's needs, and be sure that each has a good philosophical fit and compatible style with your team and business.
- It is essential that you obtain a personal introduction from another investor, entrepreneur, accountant, or lawyer.
- Phone to briefly discuss the business and determine if there are any conflicts. (Venture capitalists will not invest in competing companies as they want to focus all their assistance on just one.)
- Bring a business plan and personal references to the first meeting.
- At the first meeting, give a brief product overview of the plan and identify the key reasons for your success (this is not the time to go into detail). Tell them what is unique. What do you have to offer that others do not? Why will you succeed?
- Arrange for follow-up meetings; expect them to take place over a few weeks.

Venture Capital Directories

The preceding information suggests that you prudently and carefully present your plan to investors. First, make sure that the investment group you are approaching invests in your type of company. You do this by consulting some of the many venture capital directories and lists available. Your library should have pointers to more venture capital directories. The following are the three most useful and popular directories:

- One excellent venture capital directory, especially for those in Silicon Valley, is available for $50 and is updated yearly (usually in early May). You can get it by contacting Western Association of Venture Capitalists—Directory of Members, 3000 Sand Hill Road, Building One, Suite 190, Menlo Park, CA 94025, (415) 854-1322. Make sure you invest in the most current directory as these become obsolete very quickly.
- A national directory is also available, entitled *National Venture Capital Association Directory*. To obtain this directory, write to: NVCA, 1655 North Fort Myer Drive, Suite 700, Arlington, VA 22209.
- *Pratt's Guide to Venture Capital Sources*, updated annually, is available in some bookstores for $145, or by writing to Venture Economics, Inc., 451 Buckminister Drive, Norwood, MA 02062.

You can obtain directories in computer-readable form from a variety of sources. They claim to have database sorting capabilities that are available on both IBM- and Mac-compatible disks. Scan the latest issue of *Inc.* magazine for advertisements or new programs. The following are two useful references. (Note, however, that you will be wasting your time if you buy an electronic directory to use as a broadband mailing list to send out 100 business plans.)

- *VenCap Data Quest*, from Artificial Intelligence Research, in Mountain View, CA. A larger database ($89.95) has information on 399 venture capital firms. A smaller database ($49.95) has information on about 250 firms. Quarterly updates sell for $69 and $39, respectively.
- *The Financing Sources Databook*, from Data-Merge, in Denver, CO, contains information on about 750 venture capitalists, banks, and finance companies, and gives less complete detail on more than 2000 additional sources. The cost is $399 for the basic product. A version with fewer listings is $139. Updates are $75 and $49, respectively.

These directories will tell you:

- who the officers or partners are
- what kind of company it is and how long it has been in business
- its investment posture in terms of minimum and maximum initial investment and desired total commitment (average and maximum) over time to any one investee company

- the maturity of company desired (seed to buyouts)
- special help that can be provided in addition to venture capital
- areas of preferred investment
- areas avoided for investment

Do not begin searching for a venture capitalist without first consulting a directory.

Over the Transom

Over the transom refers to the submission of unsolicited business plans (also known as *cold deals*)—those that appear on an investor's desk without any introduction or explanation. An unsolicited business plan submitted to a venture capital firm has almost no chance of being funded.

Many entrepreneurs have wasted months sending unsolicited business plans to venture capital firms, only to get polite declines in the mail (if they hear anything at all). To make progress in getting a business plan read and taken seriously, you need either yourself or a team member to have a reputation or name recognition quality, or an introduction.

Investors, for the most part, simply will not take the time to study a plan from an unknown entity.

Introductions and How to Get Them

Getting introductions is difficult. This is not something you wait to worry about after you have finished writing your plan. You need to work on introductions long before you attempt to launch your own venture capital-backed company. Besides asking every friend you know who might know investors (your banker, doctor, associates, etc.) you need to make more friends. This is done through the process called *networking.*

Networking and Name Recognition

Attending professional association and business club meetings might seem unpleasant, especially for a technically inclined engineer or scientist. But if you want to start a venture capital-backed start-up, you must play the role of a businessperson. That means going out and meeting other businesspeople, and getting your name recognized.

Delivering a speech or a technical presentation (perhaps on the challenges of applying your technology) will go a long way in opening doors. This enables you to introduce yourself to investors. If you have given a paper at a conference, make reprints and send them to 20 or 30 potential investors with a note saying that you thought they might be interested in the topic. If you do this several times over a couple of years, your name will eventually have recognition value in their minds. An investor might not remember how he or she knows you, but your name will become familiar enough to at least glance at your business plan when it finally appears.

Looking the Part

Dress for Success is an old book, but it contains good advice. When an investor talks to you about your business idea, does he or she see an engineer or an entrepreneur? An entrepreneur knows how to sell, and you are selling yourself now. You have to offer the customer what he or she wants, not what you want him or her to want! A few years back it was somewhat classy in Silicon Valley to wear tennis shoes and try to start a computer company. It is still entertaining to incorporate fun with business, but this is only appropriate with friends.

If you obtain an interview with a potential investor, you had better look the part. For men, a modern, clean, pressed suit is mandatory, and a woman should wear a business suit. Men should wear long socks. Investors actually say such things as "Joe (the engineer) had a really good idea and a well thought-out plan, but he was wearing short socks." Little things, like a gold Cross pen in your pocket, say that you are a businessperson who should be taken seriously. Your commitment to projecting a professional appearance can yield a high return on your investment. This is not subterfuge; it is expected business behavior.

There is controversy over the "look successful—be successful" point of view. Some claim it should make no difference, but it clearly does to many investors. If you want their money, play their game. It is even suggested that you dump your 10-year-old Toyota and lease a new BMW (or at least a new Toyota). People observe how you appear personally, what you drive, and where you live. If you plan to start a business in your community in a few years, it would not hurt to be living where successful business people live when you go looking for money. Remember the saying that bankers only lend money to people who apparently do not need it? Investors often work the same way, so it helps to not look like you need their funding.

Unsolicited Business Plan

For unsolicited business plans that you submit to venture capitalists, you will seldom receive any comments to help you out, usually for one of two reasons:

- The investors did not have time to read or evaluate your plan, let alone write you a letter or talk to you on the phone.
- They do not want to risk litigation by commenting on your plan only to have you sue them when they back someone else's plan with your ideas in it.

This leads to the topic of confidentiality and nondisclosure agreements.

Confidentiality and Nondisclosure Agreements (NDAs)

Since your business plan is valuable to you and you do not want any competitors to get hold of it, you must print "confidential" on each page and treat your plan as a trade secret. You should never give your plan to anyone who does not promise (preferably in writing) that they will respect the document accordingly. This is good theory, but bad reality. Some investors will sign nondisclosure agreements (NDAs), but they are few and far between; further, they would be more comfortable with an NDA if they already knew you or the person making the introduction. If you submit an unsolicited plan to a venture capitalist with a cover letter asking him or her first to sign an NDA, the plan will likely be returned unopened. The reason, again, is the risk of lawsuits. Investors receive hundreds, even thousands, of plans each year and they cannot be expected to remember what information they saw where, or to whom they should not tell what.

It is exceedingly difficult (if not impossible) for an investor to evaluate your plan without disclosing its contents to others, and an investor is not likely to worry about putting everyone in the due diligence chain under NDA for you. The sad fact is that your business plan, especially if it is good, will probably be read and copied by many others. Most of these people will have good intentions and will not intentionally deliver your plan into the hands of a direct competitor, but it does happen.

Your best defense is not to include in your business plan your most sensitive market information or the technical aspects of building your product. Save this information for one-on-one discussions with interested investors. The purpose of a business plan is much like that of a resume: it gets you the interview. You sell and close after you get the appointment.

There is no need to overdisclose confidential information in your business plan.

Negotiation Skills

Take the time to learn negotiation skills (such as Nierenberg's win-win approach to negotiating and deal-making). Both sides should act and feel like winners in a funding agreement. Kurtzig of ASK was a big believer of leaving something on the table in negotiations. She also wrote:

> To get in or out of a deal, there are four things necessary for a successful negotiation: good sense, guts, diplomacy, and leverage. You need good sense to know what to ask for, guts to ask for what you want, diplomacy to know how to ask for it, and leverage to get it.

Paying for Criticism?

A number of firms will offer to read your business plan, make suggestions for improvements, and presumably represent you to the investment community. The back pages of business magazines and newspapers are full of these ads. Some of these firms are legitimate, but many are just out to get your money.

It is true that you may have to pay for help to write a good plan. There are many financial advisors who can do a reputable job assisting you with your financial pro formas, for example, if that is a weak area for you. Seek out specific advice as you need it, and pay for that, but do not pay big bucks for general advice. (It would be wise, too, to request that these advisors sign an NDA).

It is better, however, to get good, experienced businesspeople to work with you who will give you their money and their advice (rather than you paying them). These individuals are the angels discussed previously. Someone who is truly in a position to help to make you successful will share in the future riches he or she helps you achieve, not in your precious pre-seed funds.

Prenuptial Provisions

Venture capital ratchets are powerful instruments employed by most venture capital investors to ensure their stake in your business. A related concept (more in your favor) involves a provision in your contract with your seed and start-up investors that will ensure that they will stick with you when you need them in the future. While it is true that the investor will most likely propose the terms of any deal, everything is there for you to negotiate.

The typical scenario is that you raise some seed or start-up cash from an investor, a sort of marriage is established, and off you go on your honeymoon. But what happens later in the marriage?

As was mentioned earlier, your worst nightmare in a start-up is running out of money. What would you do if you were on plan, the time came to raise more cash, and your original investors no longer possessed the enthusiasm for your business they initially had? The marriage loses some of its passion. This situation can destroy a company, and it happens frequently enough that you need to plan in advance for it.

Individuals with significant financial assets frequently employ a prenuptial agreement before getting married. You should consider a similar agreement with your investors.

If you can negotiate it, insert a "pay to play" provision in your investment agreement that states the investor's responsibility to put in a pro rata share in future rounds of financing. If your initial investor can, and intends to, support you in subsequent rounds, put that intention in writing, and insert a penalty for a failure to perform. Penalties can be in the form of a loss of liquidation preferences or a ratchet to severely dilute and wash out shares. This provision is intended not to punish your seed and start-up supporters, but to apply financial pressure on them to set aside appropriate funds so that they can and will support you when you need them again in the future.

Software Success—Who Gets Funding How and Where?

Software Success is published monthly by marketing consultant David H. Bowen for a select readership—individuals and businesses marketing computer software. Together, through surveys and seminars, they investigate significant issues for success in the software business.

We earlier established that software is one of the easiest businesses for an engineer to start, due to low capitalization requirements, minimum manufacturing problems, and growing market interests and needs. In a recent issue of *Software Success*, Bowen reports some interesting statistics (derived from a survey of his readership) regarding funding issues for software entrepreneurs. These statistics represent an interesting slice of the real world.

Many of his readers are in the bootstrap (i.e., self-funding) state (as you might be) or are determined not to use outside money if possible. Her is a breakdown of their company revenues by percent of Bowen's total responses:

Company Revenue	Percent Responding
< $250,000	21.7%
$250,000-$500,000	15.1%
$500,000-$1 million	21.7%
$1 million-$2 million	17.8%
$2 million-$5 million	12.5%
> $5 million	12.5%

Companies with greater revenue and greater revenue potential are more interesting to investors. Private investor typically want to see a potential for $5 million in sales so that the company can be sold to a larger company later. Venture capitalists want to see an absolute minimum annual revenue potential of $25 million and many have minimum limits of $100 million.

Paid-in capital is money invested in a company for equity and the most common amount for this survey is $10,000 to $100,000. This is a common range that founders can fund from their own personal savings. Many companies understate their capital because they do not pay themselves full salaries during start-up.

The most common form of ownership was several partners (42.1%), followed by one owner (29.6%). Less than 20% of the companies who responded had outside investors.

Software companies, while easy to start, are also difficult to find funding for. Most of the companies that have less than $100,000 in paid-in capital did not raise money from outsiders. The funding came from the principles and their family and friends. The more you are willing to invest in your business, the higher venture capitalists will value your business. Investors figure that their money is safer with you if you also have your own funds tied up (or those of your friends and relatives).

For the larger companies in the survey, 15.1% of the responding companies tried to raise over $1 million, but only 11.8% have over $1 million in paid-in capital. This is pretty close to the 11.2% who had either venture or public ownership. Venture capitalists are generally interested in deals greater than $1 million.

Mid-sized companies seeking $100,000 to $1 million need too much money for family and friends to help, and too little for venture capitalists. In this range, Bowen suggests that private investors can make sense. Of the companies surveyed, 10.5% had private investors as owners. Since 18.4% of the companies had $100,000 to $1 million in paid-in capital, Bowen guesses that 57% of these situations involved private investors (10.5% divided by 18.4%).

Of the companies, 43.4% have tried to raise money from private investors, and 52% of those who tried were successful; 15% raised less money than desired, 3% gave up more stock than planned, and 12% were unsuccessful. Another 17-21% appear to have given up. While Bowen concludes that his survey suggests that it is reasonable to raise private money for software companies, you certainly do not want to bet your next payroll on obtaining private financing.

There are numerous unfortunate situations where family members invested who could not afford to lose their investment, and where private investor were later severely diluted by venture capitalists in later rounds through ratchet clauses.

In the venture capital arena, 29.6% of the companies have tried to raise venture capital. Of them, 37.8% were successful, 13% obtained less money than desired, 22.3% had to give up more stock than desired, and 26.7% were not successful. The lack of revenue does not discourage people from approaching venture capitalists, but the success rate below $5 million in company revenues is under 20% of the companies that tried. For companies over $5 million in revenue, 45% of them succeeded in raising venture capital.

On public offerings, only 4.6% of respondents even tried, but that represented 36.8% of respondents with over $5 million in revenue. Of those trying, 28.3% were successful.

Successful fund-raising takes time. The estimated average amount of time it took for successful firms to raise money was:

- one to three months for private investments
- six months to one year full time for venture capital
- more than one year to go public

Why were some companies unsuccessful in fund-raising? Too much work, did not want to give up control (the top slated reason), revenue potential no enough (probably the tope real reason), did not need enough money, and weak management team were among the reasons cited.

Growth rates relate to fund-raising success. Having a high growth rate (> 30% annually) helps somewhat to raise private money. Of companies with over 30% growth rate expectations, 56% were successful in raising private money. Only 33% of the companies with 1-10% growth rate expectations were successful. In venture capital, a high growth rate is requires. No one with expectations over 10% was successful in raising venture capital.

Bowen concludes his statement,

> Fund-raising is the one task I believe the CEO must do himself. Investors want to talk to the person at the helm. In my experience, finder are rarely successful in getting companies funded.

Edward Roberts of MIT seems to agree. In his study, he found that of all referral mechanisms leading to 54 investments in 20 high-technology firms, finders played a role in only 6 of the 54 instances. Presumably, finders are best at earning one and introduction—the entrepreneur has to do the actual selling. Finders are believed to play a more effective role in securing overseas investments. It would be difficult to secure funding from a Japanese firm, for example, without the assistance of a respected Japanese consultant on your side.

Source: Statistics in this section are excerpted from the June 1991 issue of *Software Success*, with permission of David H. Bowen.